STEPHAN KÖRNER

NIJHOFF INTERNATIONAL PHILOSOPHY SERIES

VOLUME 28

General Editor: JAN T.J. SRZEDNICKI (Contributions to Philosophy)
Editor: LYNNE M. BROUGHTON (Applying Philosophy)
Editor: STANISLAW J. SURMA (Logic and Applying Logic)

Editorial Advisory Board:

R.M. Chisholm, Brown University, Rhode Island. Mats Furberg, Göteborg University, D.A.T. Gasking, University of Melbourne, H.L.A. Hart, University College, Oxford. S. Körner, University of Bristol and Yale University. H.J. McCloskey, La Trobe University, Bundoora, Melbourne. J. Passmore, Australian National University, Canberra. A. Quinton, Trinity College, Oxford. Nathan Rotenstreich, The Hebrew University of Jerusalem. Franco Spisani, Centro Superiore di Logica e, Scienze Comparate, Bologna. S.J. Surma, Auckland University, New Zealand. R. Ziedins, Waikato University, New Zealand.

For a list of volumes in this series see final page of the volume.

Jan T.J. Srzednicki
editor

Stephan Körner – Philosophical Analysis and Reconstruction

Contributions to Philosophy

1987 **MARTINUS NIJHOFF PUBLISHERS**
a member of the KLUWER ACADEMIC PUBLISHERS GROUP
DORDRECHT / BOSTON / LANCASTER

Distributors

for the United States and Canada: Kluwer Academic Publishers, P.O. Box 358, Accord Station, Hingham, MA 02018-0358, USA
for the UK and Ireland: Kluwer Academic Publishers, MTP Press Limited, Falcon House, Queen Square, Lancaster LA1 1RN, UK
for all other countries: Kluwer Academic Publishers Group, Distribution Center, P.O. Box 322, 3300 AH Dordrecht, The Netherlands

Library of Congress Cataloging in Publication Data

```
Stephan Körner, philosophical analysis and reconstruction.

   (Nijhoff international philosophy series ; v. 28)
   Bibliography: p.
   Includes index.
   1. Körner, Stephen, 1913-    . 2. Philosophy.
I. Körner, Stephan, 1913-      II. Srzednicki,
Jan T. J.   II. Series.
B1646.K784S74  1987         192            87-12265
```

ISBN 90-247-3543-2 (this volume)
ISBN 90-247-2331-0 (series)

Copyright

© 1987 by Martinus Nijhoff Publishers, Dordrecht.

All rights reserved. No part of this publication may be reproduced, stored in a retrieval system, or transmitted in any form or by any means, mechanical, photocopying, recording, or otherwise, without the prior written permission of the publishers,
Martinus Nijhoff Publishers, P.O. Box 163, 3300 AD Dordrecht,
The Netherlands.

PRINTED IN THE NETHERLANDS

TABLE OF CONTENTS

Notes on Contributors VII

Editor's Note IX

Note on Comments and Replies XI

1. A Version of Cartesian Method 1
   RODERICK M. CHISHOLM

   Körner's Reply 9

2. Concepts, Rules and Innateness 17
   KEITH LEHRER

   Körner's Reply 27

3. Five Concepts of Freedom in Kant 35
   LEWIS WHITE BECK

   Körner's Reply 52

4. The Modes of Philosophical Involvement
   with a Categorial Framework 59
   WITOLD MARCISZEWSKI

   Körner's Reply 69

5. ESTABLISHING THE CORRESPONDENCE THEORY OF TRUTH
   AND RENDERING IT COHERENT 75
   RICHARD SYLVAN

   KÖRNER'S REPLY 84

6. PRUDENCE AND AKRASIA 89
   ROBERT A. SHARPE

   KÖRNER'S REPLY 107

7. DETERMINISM, RESPONSIBILITY AND COMPUTERS 113
   JACEK HOLOWKA

   KÖRNER'S REPLY 132

8. LOGIC AND INEXACTNESS 137
   JOHN P. CLEAVE

   KÖRNER'S COMMENT 160

BIBLIOGRAPHY OF STEPHAN KÖRNER'S WORKS 161

INDEX 169

NOTES ON CONTRIBUTORS

Lewis White Beck is Burbank Professor Emeritus of Intellectual and Moral Philosophy at the University of Rochester. His most recent book is **Kant's Latin Writings: Translations, Commentary and Notes** (1986).

* * * * * * * * * *

Roderick M.Chisholm is Professor of Philosophy at both Brown University and the University of Graz. He is the author of: **Perceiving; Theory of Knowledge; Person and Object; The First Person;** and **Brentano and Intrinsic Value**. Professor Chisholm has also been editor and translator of works by Franz Brentano.

* * * * * * * * * *

John P.Cleave is a Reader in Mathematics, University of Bristol.

* * * * * * * * * *

Jacek Holowka is a Doctor of Philosophy at the University of Warsaw.

* * * * * * * * * *

Keith Lehrer is Professor of Philosophy at the University of Arizona and Honorary Professor at Karl-Franzens Universita"t, Graz, Austria. He is the author of **Knowledge** (Clarendon Press, 1974), as well as co-author and editor of several other books. He is the subject of a **Profiles** volume - **Keith Lehrer** (Reidel, 1981), edited by R.Bogdan.

* * * * * * * * * *

Wartold Marciszewski is Professor at Warsaw University. He is head of the Department of Logic, Methodology and Philosophy of Science (the Department specializes in history of logic and in computer-aided reasoning), and is also a Board Member of the Committee for Philosophy in the Polish Academy of Sciences. One of his main interests is Leibniz's metaphysics.

* * * * * * * * * *

Robert A.Sharpe is Professor of Philosophy at Saint David's University College, University of Wales. He is the author of **Contemporary Aesthetics** (Harvester, 1983) and of many articles, principally on aesthetics, but also on philosophy psychology, philosophy of social science and on music.

* * * * * * * * * *

Richard Sylvan, an Antipodean researcher and writer resident in Australia, is trying to put together a systematic and comprehensive **deep theory**, including among other parts, deep green theory and its practice, deep pluralism and its regional applications, deep relevant logic and its dialectical elaborations, and integrated object and process theory. But he also contributes to the endless negative, often exhausting business, of criticism; in particular, he is a persistent, untireless critic of prevailing dogmatic and reductionist philosophies, such as materialism, empiricism, extensionalism, utilitarianism, economism, and, not least, Australian primitivism.

EDITOR'S NOTE

The idea behind "Contributions to Philosophy" is to present either the work of an influential philosopher, an influential school, or an influential line of thought, and to present it as it appears now. It is hoped to come as near as possible towards treating all or most of the important facets of such a contribution, so that a fair and critical picture emerges.

The conception of such a volume concerning the work of Stephan Körner arose quite some time ago. On investigating the possibilities it was discovered that Professor Adam Morton of Bristol University's Philosophy Department, had also conceived an idea of a Körner volume, but his idea was different, and not intended primarily as a commentary on Körner's work. On investigation it was decided that only one volume should be attempted at that stage, and Professor Morton took the editorship of a "Contributions" type of volume. In the event Professor Morton could not continue with the project. Thanks are due to him for the very considerable amount of time, effort and care that he contributed to this project. Professor Morton started the project and did the preliminary and initial work for it.

We have experienced some difficulty in changing the character of the volume. Due to this possible contributors were sought more than once. Those who have contributed promptly in response to the original approach had to wait rather a long time for publication, and thanks are due to them for their patience. This is also the reason why the volume did not appear at a time more appropriate to Professor Beck's dedication.

The editor and publisher wish to thank all contributors for their co-operation and the work they contributed to this book. Thanks are also due to Professor Körner himself, for his co-operation and for the replies that he produced very promptly and unstintingly.

I would also like to express my appreciation of the work of Mr.H.Imberger who worked on the proofs and checked the text; Mrs. Josie Winther who was responsible for typing and manuscript preparation; Miss Samantha Freeman who did the indexing; and Mrs. Frances Freeman who worked on the Bibliography.

J.S.

NOTE ON COMMENTS AND REPLIES

Stephan Körner

It seems proper to preface my comments and replies with the expression of my gratitude to the editor of this volume and the contributors. I have no doubt that the readers of the contributions will find them as interesting and instructive as I did. My aim in formulating the following reactions has been to deal with the philosophical issues raised rather than to defend the views which I have expressed during the past half-century. In this connection I should like to mention my book **Metaphysics: Its Structure and Function** (Cambridge, 1984), which, apart from dealing with its main topic, also contains brief indications of my present position on a variety of related problems.

I have also been honoured by another collection of essays on my philosophy, which appeared as Volume 20 of the **Grazer Philosophische Studien** (Graz, 1984) and I have, at the editor's request, written a brief reply, which will appear in due course. Fortunately, the essays contained in the two collections rarely overlap and I believe that the same is true of my replies.

1. A VERSION OF CARTESIAN METHOD

Roderick M. Chisholm

Introduction

In one of his many profound discussions of the method of philosophy, Körner makes the following observations about Descartes:

> The most remarkable claims made by Descartes are, on the one hand, that there are absolutely certain propositions which do not belong to logic or mathematics and are not based on linguistic conventions; and, on the other hand, that his method enables one to discover these absolutely certain 'truths of fact.'[1]

Körner expresses doubts about both of these claims, but I do not feel that these doubts are justified. In the present paper I shall try to set forth and defend Brentano's views. I will first discuss those "absolutely certain propositions which do not belong to logic or mathematics". And then I will discuss the method by which propositions are to be singled out. The method is a procedure that Körner himself sets forth when he describes the general nature of psychological phenomena.

Brentano and Self-Presentation*

One question that should be discussed in connection with our general topic, "Der Fundamentalismus in der Erkenntnistheorie," is this: if foundationalism is true, as I think it is, then what is the nature of the foundations of our knowledge? To this question, I think, the best answer is to be found in the writings of the Brentano school - most notably in those of Meinong and of Brentano himself concerning what Meinong had called "Selbstpräsentation," or self-presentation. And, therefore, since we are in Graz, I have chosen "Brentano and Self-Presentation" as the topic of my contribution.

---

* Introduction to the Graz talk, June 3, 1983.

RODERICK M. CHISHOLM

I will defend what Brentano says and I will suggest that even the writings of Moritz Schlick on "Das Fundament der Erkenntnis" can be understood only by reference to a doctrine such as that of Brentano.

What, then, is the doctrine of self-presentation? It is this:

Self-Presentation and Inner Perception

There are certain properties that may be said to **present themselves** to the subject who has them. These include: thinking, judging, hoping, wishing, feeling, being appeared to. If a subject has any of these properties, then, **ipso facto**, it is evident to him that he has them. These properties constitute their own evidence, so to speak. **That** the subject has them is not something that is made evident to him by anything **other** than the fact that he has them.

It was Meinong who first used the expression "to present itself" in this context. He had said, in the second edition of **Über Annahmen** (1910), that one may ascribe to certain experiences "the capability of presenting themselves, so to speak, to inner perception **(die Fähigkeit . . . sich der inneren Wahrnehmung sozusagen selbst zu präsentieren)**."[2] And he had in mind such properties as those I have cited. He had written earlier, in **Über die Erfahrungsgrundlagen unseres Wissens** (1906), that "in inner perception there is an identity between the content of apprehension and the object of apprehension **(zwischen Erkenntnisinhalt und Erkenntnisgegenstand)**.[3]

Without using the expression "Self-Presentation," Brentano had made a similar point in his **Psychology**. Thus, speaking of the sensation of hearing, he wrote: "Apart from the fact that it presents the physical phenomenon of sound, the mental act of hearing becomes at the same time its own object, taken as a whole **(Der psychische Akt des Hörens wird, abgesehen davon, dass er das physische Phänomen des Tones vorstellt, zugleich seiner Totalität nach für sich selbst Gegenstand)**.[4]

Brentano uses the expression "inner perception" instead of "self-presentation." Inner perception, he said, is characterized by its "immediate, infallible evidence **(unmittelbare, untrügliche Evidenz)**". Of all the types of knowledge of the objects of experience, inner perception alone possesses this characteristic. Consequently, when we say that mental phenomena are those which are apprehended by means of inner perception, we say that their perception is immediately evident."[5]

A VERSION OF CARTESIAN METHOD

Discovering the Directly Evident

How, then, are we to mark off those facts of inner perception which may thus be said to present themselves to the subject? How are we to know whether a particular psychical phenomenon **is** one of those phenomena that thus present themselves? We find the answer if we consider Körner's own description of psychological phenomena:

> All mental phenomena consist in a person's awareness of something - 'awareness' not being further analysable. The objects of his awareness may be either propositions, as in the case of knowledge or doubt, or else non-propositional, as in the case of desire. I shall say that a relation between a person and the object of his awareness is an 'intentional relation' and the object an 'intentional object' if, and only if, one of the two following conditions is satisfied: (i) if the object is a proposition, then the relationship between the person and the proposition does **not** logically imply that the proposition is true or that the proposition is false; (ii) if the object is non-propositional, then the relationship between the person and the (non-propositional) object does **not** logically imply that the object exists in the physical world.[6]

If a phenomenon satisfies this definition of intentionality, then, we may be sure, it is one of those phenomena that present themselves. Let us consider the application of this criterion.

Thus, if I believe that it is raining, then the fact that I do believe this is something that is immediately evident to me and requires no further investigation on my part. But my **knowing** that it is raining is not intentional - for it is necessarily the case that, if I know that is raining, then the proposition that it is raining is true. Believing that it is raining, then, is self-presenting, but knowing that it is raining is not.

Consider now an example of a different sort. Suppose I am thinking of you as being a profound philosopher. What is self-presenting in this case? Not that I am thinking about **you** - and not even that there **is** someone whom I am thinking about. What is self-presenting is merely the fact that I am thinking of someone as being of such-and-such a sort and also as being a profound philosopher.

What of **perception**? Suppose, for example, I perceive a dog. Here we may quote Körner's answer: "My **perceiving** a dog logically implies that the perceived object exists in the physical world - at least if the term 'perceiving' is used in the usual way according to which 'perceiving a physically non-existent dog' is a contradiction in terms. However...

perceiving a dog includes being under the impression of perceiving a dog and this latter relation is intentional."[7] If I perceive a dog, then it is immediately evident to me, not that I perceive a dog, but that I **take** something to be a dog. And it is also immediately evident to me that I am appeared to in a certain way - in the way in which I think that a dog would appear to me.

Similarly for other mental phenomena - hoping, wondering, remembering, liking, disliking, being pleased, being displeased. But here, too, we should remind ourselves that the self-presenting psychological phenomena are those phenomena which, as Körner puts it, imply that fact of awareness. Psychological phenomena that are merely dispositional - being irascible, say - will not be self-presenting. How are we to tell, then, whether a psychological phenomenon is merely dispositional?

Consider this example. If you are my friend, is it evident to me that I like you? If I am now enjoying your company, then the fact that I am enjoying someone's company is now evident to me. But it is not thereby evident to me that I am related to you as a friend. It is not evident to me, say, that I will try to act in a way that will please you or that is to your advantage, or even that I am disposed to act in such a way. But what is evident to me is the way that I feel at the moment.

I would say, then, that we do have a method for singling out those "absolutely certain propositions which do not belong to logic or mathematics". They are those propositions satisfying the criteria of intentionality that Körner has set forth.

And there is still **another** way of singling out these same propositions. This was proposed by Moritz Schlick when he set forth his criterion for singling out what he had called (somewhat misleadingly) "observation statements," and also "confirming statements." Schlick wrote:

> While in the case of all other synthetic statements determining the meaning is separate from, distinguishable from, determining the truth, in the case of observation statements they coincide, just as in the case of analytic statements. However different therefore "confirmations" are from analytic statements, they have in common that the occasion of understanding them is the same as that of verifying them [**der Vorgang des Verstehens zugleich der Vorgang der Verifikation ist**]: I grasp their meaning at the same time as I grasp their truth [**mit dem Sinne erfasse ich zugleich die Wahrheit**]. In the case of confirmation it makes as little sense to ask whether I might be deceived regarding its truth as [it does] in the case of a tautology. Both are absolutely valid. How-

ever, while the analytic, tautological statement is empty of content, the observation statement supplies us with the satisfaction of genuine knowledge of reality.[8]

It is interesting to note that Schlick never cites any plausible examples of his "confirming statements." But is clear that, had he done so, the examples would also have illustrated "self-presentation," or "inner perception."

But there is another feature of Brentano's view that Schlick would not have found acceptable. This is the view that the **subject** of experience is included in every presentation.

Primary and Secondary Objects of Consciousness

Meinong's term "self-presentation" suggests two possible interpretations: presentation of **itself** and presentation of **the** self. Both interpretations are justified. To see this, we should consider the distinction Brentano makes in the **Psychology** between the **primary** and the **secondary** objects of intentional phenomena.

Taking the act of hearing as an example of a sensation, Brentano writes: "The act of hearing appears to be directed toward sound in the most proper sense of the term, and because of this it seems to apprehend itself incidentally and as something additional."[9] He refers to the sound as the **primary object** of the act of hearing and the act of hearing itself as the **secondary** object.[10] This distinction applies to every sensation and to **presentation** [**Vorstellung**] in general.

Aristotle had put forth a similar view in the **Metaphysics**. He wrote: "Knowledge, sensation, opinion and reflection seem always to relate to something else, and only incidentally to themselves."[11] Here Aristotle is concerned to say that the relation that these phenomena bear to ourselves is **only** incidental; but the important point is the implication that these phenomena in fact always do relate to ourselves, even if only incidentally.

Brentano wrote: "The presentation [**Vorstellung**] of the sound and the presentation of the presentation [**die Vorstellung von der Vorstellung**] of the sound form a single mental phenomenon; it is only by considering it in its relation to two different objects, one of which is a physical phenomenon and the other a mental phenomenon, that we divide it conceptually into two presentations. In the same mental phenomenon in which the sound is present to our minds we simultaneously grasp [**erfassen**] the mental phenomenon itself. What is more, we grasp it in accordance with its dual nature insofar as it has the sound as content within it, and insofar as it has itself as content at the same time [**insofern es als Inhalt den Ton in sich hat, und insofern es zugleich sich selbst als Inhalt gegenwärtig ist**]".[12]

Brentano sometimes says, not that the **act** is the secondary object of presentation, but that the **thinker** is the secondary object of presentation. This way of putting the matter fits best with Brentano's final views. We should remind ourselves that he is a reist. He believes that true statements that are ostensibly about the **experiences** of a person can be reduced to statements about those concreta that he calls the **accidents** of the person. According to this view, statements ostensibly about the **act** of thinking are reducible to statements about the **thinker** himself. For example, the apprehension of **someone hearing a sound** is reduced to the apprehension of **the subject as hearing a sound**. Where others would speak of the **occurrence** of that state of affairs which is the hearing of a sound, Brentano speaks of the **existence** of that accident which is the subject hearing a sound. Hence we could say, not that the **hearing** of the sound is the secondary object, but that the **hearer** of the sound is the secondary object.

The primary object of a presentation, then, is something other than that presentation. The primary object of **hearing** is the thing that is heard. And the secondary object is the presentation itself or, better, **the subject as having that presentation**. Hence we may say that the secondary object of hearing is the hearer as having the primary object. We must not suppose that there is one act of presentation that presents the primary object and a **second** act of presentation that presents the secondary object. There is one presentation: that of the subject experiencing the primary object. This presentation is its own secondary object; and the primary object, we might say, follows from the secondary object.

Hence we can see how the self may be said to be a part of every presentation. And so the ambiguity of Meinong's term "self-presentation" has a justification: inner perception presents **itself** and it also presents **the** self. If I experience a red sensation, then the presentation is **my having a red sensation**. It is its own secondary object and has the red sensation as a primary object. This theory of presentation does not lead to an infinite regress. Brentano is not telling us that every presentation may be extended to a wider presentation that has the first presentation as its object. We don't get the experience of the subject by **adding** something to the primary object of presentation; rather, we get the primary object of the presentation by **subtracting** something from the experience of the subject. The supposed regress, as Brentano tells us, "ends with the second member."[13]

Here, then, we have a plausible defence of the view that there are "absolutely certain 'truths of fact'" and that these truths imply the existence of the psychological subject.

* * * * * * * * * *

FOOTNOTES

1. Stephen Körner, **Fundamental Questions of Philosophy** (Sussex, Harvester Press, 1979), p.23. Körner recently made a similar point in a lecture, "Über philosophische Methoden and Argumente," presented to a colloquium on "Fundamentalismus in der Erkenntnistheorie," at the University of Graz, June 4, 1983.

2. **Über Annahmen,** 2nd edition, p.138; Band IV of **Meinong Gesamtausgabe,** (ed.) Rudolf Haller (Graz, Akademische Druck - u.Verlagsanstalt, 1977).

3. **Meinong Gesamtausgabe** Band V, p.432.

4. **Psychology from an Empirical Standpoint** (London, Routledge & Kegan Paul, 1973), p.129 (the published translation has "object and content" in place of "object"); **Psychologie vom empirischen Standpunkt** Band I (Hamburg, Felix Meiner Verlag, 1973), p.182. Meinong noted that he could accept Brentano's account only with reservations. He uses the term "self-presentation" in the first chapter of **Über emotionale Präsentation** (1917) and there sets forth his reservations about Brentano's view. See Band III of the **Meinong Gesamtausgabe,** (ed.) Rudolf Kindinger; translated into English as **On Emotional Presentation,** by Marie-Luise Schubert-Kalsi (Evanston Ill., Northwestern University Press, 1972).

5. **ibid.**

6. **Fundamental Questions of Philosophy,** pp.92-93.

7. **op.cit.,** p.93.

8. Moritz Schlick, "The Foundations of Knowledge," in **Logical Positivism** (ed.) A.J.Ayer (Glencoe, The Free Press, 1959) pp.209-227; the quotation appears on p.225. The original version of the paper, "Über das Fundament der Erkenntnis," may be found in Schlick's **Gesammelte Aufsätze** (Vienna, Gerold and Company., 1938) pp.289-310; the German version of the quotation appears on pp.308-309.

9. **Psychology,** p.128, my italics. The German reads: "Dem Tone erscheint das Hören im eigentlichsten Sinne zugewandt, und indem es dieses ist, scheint es sich selbst nebenbei und als Zugabe mit zu erfassen" (**Psychologie** Band I, p.180).

10. **Psychology,** p.132.

11. **Metaphysics,** 1074b, pp.35-36 (quoted in the **Psychology,** p.132). Compare **De Anima,** III, 2: "It is through sense that we are aware that we are seeing or hearing."

12. **Psychology,** p.127; **Psychologie** Band I, pp.179-180. Another succinct statement may be found in **Religion und Philosophie,** p.226, but evidently it was written by Kastil and not by Brentano (see p.265).

13. **Psychology,** p.130.

* * * * * * * * * *

REPLY TO PROFESSOR CHISHOLM

Stephan Körner

As often before in our conversations, discussions and cooperation (as coeditors of a volume of Brentano's writings), Professor Chisholm has forced me to rethink and to clarify my position on an important philosophical issue. And as always, I am deeply grateful for his efforts. Chisholm finds an apparent inconsistency between on the one hand my rejection of Descartes' method on the ground that it is intended to provide a criterion of "absolutely certain 'truths of fact'" and on the other hand my "description of the general nature of psychological phenomena". For, Chisholm argues, this description is not only, as we both agree, indebted to Brentano's psychology but is - or could in the spirit of that psychology be - expressed by absolutely certain factual propositions.

My reply falls into two parts: The first contains an examination of the notions of a truth of fact and a truth of reason, as well as an attempt at analyzing the notion of an absolutely certain proposition, in particular an absolutely certain factual (as opposed to logically necessary) proposition (Sect.1). The second section contains an argument to the effect that neither my characterization of psychological phenomena, which is not intended to result in absolutely certain propositions, nor Brentano's, which is so intended, do result in such propositions (Sect.2).

1. On Truths of Fact, Truths of Logic and the Problem of Absolute Certainty

By a "truth of fact" I understand, in accordance with Leibniz and Brentano, a proposition which can be denied without logical inconsistency and which correctly characterizes a phenomenon or class of phenomena given in inner or outer perception. A truth of fact is thus synthetic and empirical or **a posteriori**. By a "truth of logic" I shall correspondingly understand a proposition which is analytic, i.e. a proposition the truth of which is evident **ex terminis** and which, consequently, cannot be denied without internal contradiction. In drawing the distinction between truths of fact and truths of logic one raises a number of issues of philosophical importance. Among them is the question whether and, if so, how the decision as to whether a proposition is a truth of logic is dependent on the acceptance of a certain logic, rather than one of its alternatives. Another question, raised by the distinction, is the question whether there are synthetic propositions which are not empirical but **a priori** in Kant's or some other sense of the term. Although I have tried to answer both questions, they need not be considered here, since they fall outside the range of the issues, raised by Chisholm.[1]

A truth of fact must be distinguished from the fact which it correctly characterizes - even if this fact involves a proposition as a component. Thus, that somebody believes a certain proposition, is not a proposition but a fact involving a proposition as one of its components. A further, perhaps less obvious, distinction must be made between the successful identification of a fact by a proposition and the correct characterization of the fact by the proposition. For it makes good sense to say that the same fact or class of facts is successfully identified by two mutually inconsistent propositions, but correctly characterized by at most one of them. Examples are the identification and characterization of freely falling objects by the Newtonian and Einsteinian dynamics or the identification and characterization of psychological phenomena by Brentano and Meinong.

Although Descartes does not distinguish between the various types of propositions the absolute certainty of which his method is supposed to establish, he does try to establish the absolute certainty of synthetic propositions which are true of inner perception. An example is the proposition 'I possess a spirit which is different from my body', the absolute certainty of which is based on its "necessary connection" with the proposition 'I know something'.[2] My rejection of Descartes' method for establishing absolutely certain propositions of fact or indeed of any other absolutely certain propositions, is, as I explain elsewhere in some detail,[3] based on my finding his examples of intuitively certain propositions unconvincing and his concept of necessary connection a self-contradictory or, at least, obscure attempt to combine the logical necessity of deductive inference with the ampliative nature of non-deductive inference.

REPLY TO CHISHOLM

The account of absolutely certain propositions given by Brentano is free from such obscurity. He admits and distinguishes between two kinds of absolutely certain propositions: those which are absolutely certain because their truth follows **ex terminis** from their logical analysis; and those which are absolutely certain because they describe some part or aspect of inner perception. Yet his notion of description, which is fundamental to his position, stands in need of logical analysis. My purpose in attempting it here, is on the one hand to refute Chisholm's claim that my account of psychological phenomena implies that the propositions characterizing them are absolutely certain; on the other hand to prepare the ground for a critique of Brentano's descriptive psychology.

The analysis of the notion of description is based on contrasting 'descriptive concepts' with 'interpretative concepts' or, more precisely, on distinguishing between concepts of different interpretative levels.[4] Let $P(x)$ and $Q(x)$ be two perceptually applicable concepts, i.e. concepts which are applicable to something that is given in perception. $P(x)$ is then defined as interpretative of $Q(x)$ if, and only if, (1) $P(x)$ and $Q(x)$ do not differ perceptually or, more briefly, are coostensive; (2) $P(x)$ logically implies $Q(x)$, but $Q(x)$ does not logically imply $P(x)$ - more precisely, the applicability of $P(x)$ to an object logically implies the applicability of $Q(x)$ to it, but the converse logical implication does not hold. If $P(x)$ is interpretative of $Q(x)$, then there is, or may be made, available a concept $A(x)$ such that $P(x)$ logically implies, and is logically implied by, the conjunction of $Q(x)$ and $A(x)$. $A(x)$ may be called the non-perceptual difference between $P(x)$ and $Q(x)$. If $P(x)$ is interpretative of $Q(x)$ and of $R(x)$ and if $Q(x)$ is interpretative of $R(x)$ then $P(x)$ is of higher interpretative level that $R(x)$.

Examples of the so defined concepts can be easily given. Thus let $P(x)$ = 'x is a chair in the sense of Plato's theory of Forms'. $Q(x)$ = 'x is a chair as conceived by non-philosophical commonsense' and $R(x)$ = 'x seems to me to be a chair (though it may turn out to be an illusion)'. $P(x)$ is interpretative of $Q(x)$ and of $R(x)$ and of higher interpretative level than $Q(x)$. The non-perceptual difference between $P(x)$ and $Q(x)$ is 'x participates in a Platonic Form (after the fashion of commonsense material objects rather than perceptions which may or may not be illusory)'. There is no need here for more elaborate examples, or for exhibiting the fundamental role of interpretative concepts in commonsense, scientific and philosophical thinking.

In terms of the so far explained notions others can be defined, of which the notion of a descriptive or, more precisely, a **purely** descriptive concept, is in the present context particularly important: A perceptually applicable concept $D(x)$ is a purely descriptive concept if, and only if, it is (1) not interpretative of any perceptually applicable concept which is, or could be made, available and (2) is not the non-perceptual difference between two concepts one of which is interpretative of the other. If $D(x)$ is a purely descriptive concept, then it may, of course, be interpreted by a variety of interpretative concepts, including concepts which are exclusive of each other. On the other hand, if $D(x)$ is purely descriptive then its correct application to a fact or other entity is absolutely certain in the sense that, not being interpretative, it **a fortiori** does not admit of an alternative interpretation. It may be worth emphasizing that a concept used to characterize a phenomenon which is in some sense self-evident, is not therefore, purely descriptive and that the proposition which is asserted when the concept is applied to the phenomenon need not be absolutely certain. Though it may be self-evident to me that I have an experience which I correctly or incorrectly characterize as a pain, the self-evidence of the experience, does not imply that my concept 'x is a pain' is purely descriptive.

The interpretative character of a perceptually applicable concept can be definitely established by indicating a concept of which it is interpretative or by showing that it logically implies a non-perceptual component (e.g. 'x is caused', 'x is a substance in Kant's or Aristotle's sense'). In trying to establish the purely descriptive character of a concept, one can do no more than point out that so far no concept is available of which it is interpretative or that no non-perceptual concept is available which is logically implied by it. It may indeed be doubtful - and I am inclined to doubt - whether there are any purely descriptive concepts. However, no answer is needed in support of the following arguments, namely that my characterization of psychological phenomena neither consists in the application of purely descriptive concepts nor is intended to consist in the application of such concepts; and that Brentano's characterization of psychological phenomena, contrary to his intention, also does not consist in the application of purely descriptive concepts.

2. Psychological Phenomena and Absolutely Certain Propositions

My characterization of mental or psychological phenomena is, as I say in the sentence immediately preceding the passage quoted by Chisholm "greatly indebted to Brentano" although I would not expect it "to be acceptable to him". A strong reason for this conjecture is that I regard neither his nor my characterization of psychological phenomena as purely descriptive, i.e. as consisting in the application to them of purely descriptive concepts. As to my characterization, it contains concepts of a high interpretative level, such as the concepts of person, proposition and relation. Because of their high interpretative level, these concepts are likely to differ from corresponding concepts of some of my readers. For the purpose of the characterization of mental phenomena we can of course ignore such differences, e.g. whether being a person does or does not logically imply having an immortal soul, whether being a proposition does or does not logically imply being expressible in a language; whether being a relation does or does not logically imply being reducible to a conjunction of monadic predicates. The irrelevance of these and similar differences may be expressed by the requirement of reducing all the concepts occuring in the characterization of mental phenomena to the lowest possible level, i.e. the requirement of reducing them to their uninterpretative, purely descriptive core. We can postulate the realizability of such a "phenomenological reduction", but - as has been pointed out earlier - have no test of its having been realized.

My characterization of mental phenomena is intended to consist in a correct application to them of interpretative concepts (e.g. 'person', 'proposition', 'relation') and to remain correct if these interpretative concepts are replaced by different interpretative concepts expressed by the same words (e.g. the English words "person", "proposition", "relation"). It thus lies between two extreme methods of classification, namely a mere identification of phenomena, which may be successful even if it involves a misapplication of concepts, and a pure description through the application of purely descriptive concepts.

While Brentano's characterization of mental phenomena is meant to be purely descriptive, i.e. to consist in the application of purely descriptive concepts and, hence, to yield absolutely certain propositions of fact, it does not achieve this aim. For, it is easily shown that his concept 'x is a psychological phenomenon' is interpretative or, what comes to the same, logically implies at least one non-perceptual component. The non-perceptual component in question is the concept 'x causes y'. There are many passages in Brentano's works according to which 'x is a psychological phenomenon' logically implies 'x is caused' or, more precisely, 'There is a y such that x is caused by y'. Here it will suffice to draw attention to two good reasons for interpreting Brentano's philosophy as containing this implication. One is that Brentano regarded the principle of causality as true **ex terminis**. The other is Brentano's division of psychology into a descriptive part, which describes psychological phenomena and an explanatory part which exhibits their causes.[5]

That the concept of causality is not merely descriptive of what is given in perception, is widely agreed, although for different reasons. Thus Hume holds that 'x causes y' is not descriptive of what is subjectively given, because it is not a concept at all - being neither an impression nor an idea, but based on confusing a subjective expectation with an external necessitation. Kant holds that 'x causes y' is not descriptive of what is subjectively given, because its being applied to what is so given interprets it as objective or confers objectivity on it. Indeed Kant holds - in my view mistakenly - that an objective experience which is not causally determined is impossible. Brentano's view is even more radical because he holds, as has been mentioned above, that the principle of causality is true **ex terminis** or "is a special case of the universal law of contradiction".[6]

That this is a mistake, can be seen by comparing any two formulations of the principle of contradiction and the principle of causality e.g. (1) No proposition is both true and false and (2) 'x is an event' (or more generally, 'x is something that is becoming in time', **'ein Werdendes'**) implies 'x is causally determined' (and e.g. not merely probabilified). If we deny (1), then every proposition can be proved to be true. If we deny (2), this is not so, **unless** we assume that there is only one concept of event and, hence, only one correct analysis of what is meant by "event". This assumption of exclusiveness is, **inter alia** explicitly made in the last paragraph of **Die Lehre vom Richtigen Urteil**.[7] It follows from Brentano's mistaken assumption that the concept of event which he analyses as causally determined, is purely descriptive. Thus Brentano is right in asserting the (analytic) proposition that **his** concept 'x is an event' logically implies 'x is caused'. And he is wrong in implying the (synthetic) proposition that his concept of event does not admit the possibility of alternative interpretative concepts which, though coostensive with it, differ from it in logical content. Since the conjunction of the two propositions is false, it is **a fortiori** not absolutely certain.

Having tried to show that neither Brentano's characterization of psychological phenomena nor mine results in absolutely descriptive propositions and that, consequently, neither characterization results in "absolutely certain truths of fact", I return to Chisholm's charge of inconsistency and plead that it be changed to a charge of insufficient clarity. To this charge I plead guilty, but hope that I have made some amends in the preceding remarks.

* * * * * * * * * *

FOOTNOTES

1. See e.g. **Fundamental Questions of Philosophy** Chapter 3 and Chapter 12.

2. **Rules for the Direction of the Mind, Comments on Rule XII.**

3. See the writings mentioned by Chisholm.

4. For details see e.g. **Conceptual Thinking** (Cambridge, 1955) pp.138ff. and **Categorial Frameworks** (Oxford, 1970) pp.51ff.

5. See e.g. Alfred Kastil's excellent introduction to Brentano's philosophy **Die Philosophie Franz Brentano's** (Bern 1951) p.22, p.29 et **passim**.

6. See **op.cit.** p.22.

7. Edited by Franziska Mayer-Hillebrand from Brentano's literary bequest (Bern, 1956).

* * * * * * * * * *

Srzednick, J. T. J., Stephan Körner – Philosophical Analysis and Reconstruction. ISBN 90-247-3543-2.
© *1987. Martinus Nijhoff Publishers, Dordrecht. Printed in The Netherlands.*

2. CONCEPTS, RULES AND INNATENESS

KEITH LEHRER

Professor Körner formulated a theory of ostensive concepts in his book, **Conceptual Thinking**, which remains a good summary of his views. When I was only a student of philosophy, I met Professor Körner at Brown University and questioned him about the matter of ostensive rules which he used to explicate the notion of ostensive concepts. At the time, he asked me what theory I would propose in place of his. I have thought about that now and then through the years, and I should now like to consider an answer, albeit, one that Thomas Reid published in 1764 in **An Inquiry into the Human Mind on the Principles of Common Sense**. This answer is in terms of innate conceptual principles. I claim no originality either for the objections to the sort of theory Professor Körner defends or the solution, which is due to Reid, but I think that the objections are fundamental issues in cognitive psychology and the philosophy of language. As a result, the issue seems to me to be of importance, and I should like to use this occasion to provoke Professor Körner to reply. In fact, I think that the proposal I shall make may well be a consistent modification of his theory rather than a systematic alternative to it. Whether we agree or disagree, his reflections will, I believe, be of interest to many philosophers beside myself.

The critical question concerns these remarks from Professor Körner's book:

> We have argued that the meaning of the term "conceptual thinking" is ultimately given by examples, and typical examples of it are the application and acceptance of ostensive rules. We have, moreover, assumed as an empirical fact, which we are not concerned to establish, that all conceptual thinkers accept ostensive rules and thus ostensive concepts. This fact suggests a minimal definition of 'conceptual thinker' as 'accepter of ostensive concepts'. It also lends plausibility to a further assumption concerning concepts, namely, that the acceptance and application of ostensive ones is in some way fundamental to the use of all other concepts. (Körner, 31)

He says of ostensive rules,

> As has been said, the most familiar way of formulating an ostensive rule is to say, with appropriate pointing gestures, that this and this and this, and everything like it is to be assigned a certain label. (Körner, 32)

Such labels are ostensive predicates. The relation between such predicates and ostensive concepts is formulated, presupposing a rule for ascertaining synonymity, as follows:

> If we are not free to substitute for the sign any of its synonyms that we please, then we are using the sign as a predicate. If we are so free we are using it as a concept. (Körner, 14)

He remarks,

> For our purposes everybody might limit himself to the consideration of ostensive rules formulated by himself for himself. (Körner, 33)

Earlier, attempting to explain what is involved in **accepting** a rule he says,

> A person accepts a rule if he satisfies it intentionally - or at least has the intention to satisfy it. Of the presence of this intention there are various more or less reliable indications. Thus it may be evidence of the intention if the accepter can formulate his intention (and therefore also formulate the rule which he intentionally satisfies or intends to satisfy). The intention to satisfy a rule may also be indicated by less impressive evidence, for example, by the fact that on the occasion of the breaking of the rule the accepter of it notes, however vaguely, that some rules has been broken. (Körner, 6)

The basic idea is clear. To understand an ostensive concept one must accept an ostensive rule. There are minor objections. The most important is that things may resemble a number of objects in various respects. Consequently, in order to use such a rule, for example, for applying the predicate "green", one must have a certain respect in mind. But to have such a respect in mind one would already have to possess an ostensive concept characterizing the respect in which the items are alike, for example, that they are alike in respect to color. If color is an ostensive concept, as it seems to be, one would need to have another rule, an ostensive rule, this time one for the word "color" and a regress threatens.

It is another regress argument that I think is of the greatest current interest. It was formulated rather a long time ago, before Körner's book was written, in fact, by Sellars. It is based on the assumption that if understanding a predicate is to accept a rule, then learning to understand a predicate would lead to a regress, according to Sellars, as follows:

> **Thesis.** Learning to use a language (L) is learning to obey the rules of L.
> **But,** a rule which enjoins the doing of an action (A) is a sentence in language which contains an expression for A.
> **Hence,** a rule which enjoins the using of a linguistic expression (E) is a sentence in a language which contains an expression for E - in other words, a sentence in a **meta**language.
> **Consequently,** learning to obey the rules for L presupposes the ability to use the metalanguage (ML) in which the rules for L are formulated.
> **So that** learning to use a language (L) presupposes having learned to use a metalanguage (ML). And by the same token, having learned to use ML presupposes having learned to use a **meta**metalanguage and so on.
> **But** this is impossible (a vicious regress).
> **Therefore,** the thesis is absurd and must be rejected.

Of course, the **thesis** is not one explicitly advocated by Körner, but it is not difficult to supply the argument from Körner's philosophy to it. The argument is that if using a predicate or concept is to accept a rule, then learning to use a predicate or concept is to learn to accept, that is, obey, a rule. (Sellars, 321)

As Sellars quickly notes, behavior can **conform to a rule** without the person having the intention of doing so, but such conformity can be merely accidental and does not suffice for understanding the word or concept involved. Sellars proposes that there is a distinction to be drawn between mere accidental conformity to a rule and pattern governed behavior, which, though it does not involve the intention to obey a rule suffices for understanding a concept. It is not clear, however, how conformity to the most elaborate patterns of linguistic response will account for what it is to understand a word or concept. A very complex pattern of responses may not be accidental, especially if the pattern realizes some end or purpose. However, if mere conformity to a rule conceived of as making a specific response to a specific stimulus in specific circumstances does not suffice for understanding a concept, then the most complex pattern of responses will not suffice either. This is really a corollary of the fact that behavior underdetermines mental states, the very fact that refuted logical behaviorism. Thus, if a stimulus S produces a response R in circumstances C in an organism because the behavior of the organism is governed by a rule belonging to some complex pattern, we can, in principle, articulate an alternative theory in which the responses to stimulations in such circumstances result from the behavior being governed by a different rule belonging to some other complex pattern. If the rules in question merely take us from a specified behavioral input in specified circumstances to a specified behavioral output, then no matter how complex the pattern by which this connection is achieved, and no matter what ends it may realize, we may account for the same connection in terms of another rule belonging to a different pattern. This is to say that when we observe such behavior we may attribute something to the organism, pattern governance perhaps, but that does not suffice to yield the result that the organism understands the concept. If it did, logical behaviorism would be tenable, which it is not. I am not sure what Sellars would reply to this objection. I would guess that it is his view that behavior being governed by a rule according to a pattern explains but does not suffice for understanding a word or concept.

The point about pattern governed behavior also applies to some cruder forms of functionalism. They assume internal programming, but when they turn to explain conceptual content of the internal code, they appeal to responses to perceptual stimulation. The model that results is one in which sensory stimulation is processed, and, as a result of processing in terms of the program encoded in the machine language, we obtain responses. The attempt to give a functional account of content in terms of stimulus and response patterns falls victim to the same argument concerning the relation between understanding a concept and pattern governed behavior. The input and output underdetermine understanding and cannot account for it. We may put the matter more formally. Functionalism gives us a complex pattern of functions on functions, functions from sensory input to internal encoding which involves yet other functions to articulate syntax and other features of language. But functions on functions are reducible to simple functions. That is a logical fact. And a function from a stimulus in specified circumstances to a response underdetermines content and is, therefore, something less than the understanding of a concept. Something is clearly missing from the picture presented by all such accounts, something that necessitates the attribution of understanding to humans. It is, I suspect, for this reason that Körner introduces the notion that the person have the **intention** to follow the rule when he explains what is involved in accepting a rule. The problem with this proposal, however, is that it appears to lead to the regress formulated above.

More recently, Fodor has argued that if learning to understand predicates involves learning to understand that the extension of the predicate falls under certain rules, which it does, then we must have an innate language in which to represent the rules. He argues as follows:

> Learning a language ... involves learning what the predicates of a language mean. Learning what the predicates of a language mean involves learning a determination of the extension of those predicates. Learning a determination of the extension of the predicates involves learning that they fall under certain rules (i.e. truth rules). But one cannot learn that P falls under R unless one has a language. In particular, one cannot learn a first language unless one already has a system capable of representing the predicates in that language **and their extensions**. And, on pain of circularity, that system cannot be the language that is being learned. But first languages **are** learned. Hence, at least some cognitive operations are carried out in languages other than natural languages. (Fodor, 63-64)

This argument would be better formulated as a regress argument showing, as Sellars argued, that to learn a language one would need to posit a metalanguage in which to formulate the rules of the language, and to learn that language one would have to posit a metametalanguage to formulate the rules of that language, and so on. Fodor argues that to avoid the problem of circularity or regress, we must posit an internal language of thought that is innate to the organism.

Thomas Reid had given a similar argument for innate or what he called **original** language which he contrasted with conventional or what he called **artificial** language. He argued,

> I think it is demonstrable, that, if mankind had not an (original) language, they could never have invented one by their reason and ingenuity. For all (artificial) language supposes some compact or agreement to affix a certain meaning to certain signs; but there can be no compact or agreement without signs, nor without language; and, therefore, there must be (original) language before any (artificial) language can be invented... (CS,32)

He goes on to claim more specifically that in order to learn the connection between the sounds others make and what they signify that "... we discover this connection by experience; but not without the aid of (original) language." (CS, 91) He adds, "When we begin to learn our mother tongue, we perceive, by the help of (original) language, that they who speak to us use certain sounds to express certain things... ."(CS, 92) Thus, Reid thought it evident that to discover the rules of artificial language, particularly rules articulating the meaning of words, we must have an original language in which to formulate those rules. Therefore, acceptance of rules, which Körner admits to be essential to conceptual thinking, presupposes a language, as Reid put it, an original language. Reid proposed that the signs of our natural language include the "features of the face, gestures of the body, and modulations of the voice." (CS, 90)

The argument of Reid and Fodor results from a dilemma. The assumption that understanding a word or concept involves the intention to follow a rule leads to a regress, but mere conformity to a rule is insufficient to warrant attribution of understanding. What is another alternative? It was, as I noted above, sketched by Reid in 1764. Reid's thesis was twofold. Reid claimed that there were principles of the human mind, first principles, that took us from sensation resulting from sensory stimulation to the conception of some quality as well as an object having that quality and, in ordinary circumstances, to conviction of the existence of the quality and object. According to Reid, this conception was due to a faculty of the human mind, the faculty of perception, and did not involve reasoning of any kind. Reid regarded the mind as involved in active operations rather than merely passive effects and thus anticipated the computational theory of mind. Thus, according to Reid, the possession of certain conceptions need not involve the intention to follow any rule or even the first principles he mentions. On the contrary, the conception arises quite automatically. Without any intention to follow a rule, we follow it because it is our nature to do so, and we cannot really do otherwise. These conceptions arise from innate principles of our natural constitution.

Reid does not hold the view that all empirical concepts are innate or arise in this way, only those of the primary qualities. It is the primary qualities that are signified by the sensations which are natural signs of those qualities because of our constitution. His theory of our concepts of secondary qualities is that they are relative conceptions of some quality that is the cause of the sensation we have. In the case of primary qualities, however, Reid maintained that we pass from sensation to conception of the quality quite automatically and that our conception is clear and distinct.

I should like to undertake the defense of Reid's theory, but I shall reserve that task for another place. What is crucial so far is Reid's notion that at least some of our conceptions, some empirical concepts, are and must be the result of innate principles of the human mind. In this part of his theory there is some resemblance to Fodor's, for Reid defends what I think may be regarded as the computational theory of mind and the notion that there are concepts that are innately determined.

There is, however, a difference between Reid and contemporary theories that is relevant to the issue of the attribution of the understanding of concepts. For Reid the idea that the ascription of belief, conception and understanding is a **theory**, a theory of folk psychology, which competes with other theories of the human mind, is absurd. The existence of these operations of the mind is not a theory but a fact. To reject what is now called 'folk psychology' for the sake of some scientific theory would be to reject the facts for the sake of theory. With what justification does Reid regard the existence of sensation, conception, belief and so forth as fact? The argument is a simple one. Our conception of these states results from a first principle of a faculty of the human mind, namely, the faculty of consciousness. When operations of our mind such as sensation, conception and belief occur, though we may not direct our attention to them, we automatically have a conception of them from consciousness and are convinced that they exist. Thus, according to Reid, it is not from education or inculcation that we believe that we have sensations, that we have conceptions, that we have convictions; it is from our very nature, from our natural constitution. By the faculty of consciousness we have conception and conviction of the existence of these things, and by reflection we can improve and refine our observation of them. Reid's argument for the innateness of such principles was that we are powerless to throw them off and that all men have such conceptions. All men testify to the existence of sensations, conceptions and convictions, excepting, as Reid notes, lunatics and some philosophers.

Now, however, we face an important question, to wit, what reason have we to think that the operations of the human mind, the operation of our faculties, give us the truth? Reid's answer, found in his later work, is ingenious and important. It is a dilemma. Either we assume that our faculties are trustworthy or we do not. If we do not assume that our faculties are trustworthy, then we have no way of finding the truth since we must use our faculties to find the truth. If we do assume that our faculties are trustworthy, then we have initially no reason to prefer one faculty to another.

Reid contends that in addition to first principles of perception, memory, consciousness and our other faculties, it is a first principle that our faculties are not fallacious. He says,

> Another principle is -- **That the natural faculties, by which we distinguish truth from error, are not fallacious.** If any man should demand a proof of this, it is impossible to satisfy him. ...because, to judge of demonstration, a man must trust his faculties, and take for granted the very thing in question. (IP,275)

Thus we have no reason to prefer reason to perception, or either to consciousness. The faculty of perception tells us that we have bodies that move. The faculty of consciousness tells us that we have minds that think. In neither case are we confronted with a theory or a hypothesis that is the fabrication of reason but with conceptions and convictions arising from our natural constitution. The assumption that we think and feel is no more a hypothesis than is the assumption that we are live and move. These are facts that we know from our human constitution. We do not infer these things, they are natural to us, and we are powerless not to believe them.

I take it to be an empirical matter exactly what we should posit as innate mechanisms producing concepts in order to explain the conceptual understanding we actually possess. What I take to be philosophically interesting are the two claims, firstly, that our knowledge that people think conceptually depends essentially on the evidence of consciousness, and, secondly, that to account for such conceptual thinking we must assume the existence of at least some innate mechanisms of conceptual thinking. The point that is important to a discussion of Körner is that to have an intention to follow a rule presupposes sophisticated conceptual thought and, therefore, that one cannot explicate all conceptual thought in terms of the **intention** to follow a rule. It may be that Körner agrees with this, for in the passages quoted above he says only that such intentions are sufficient for accepting a rule not that they are necessary. What we need, therefore, is some explanation of what conditions are necessary that does not run afoul of the regress and circularity arguments of Sellars, Fodor and Reid.

I should like to conclude by suggesting a hypothesis, also proposed by Reid, that would enable us to reconcile Körner's views on ostensive concepts with Reid's and thereby avoid the regress and related problems discussed above. Reid thought that there were principles of our faculties that could function in two different ways. Consider some sensation, a visual one perhaps, that is the sign of motion, of a moving object. Now, according to Reid, the sensation is a **natural** sign in the sense that there is a principle of our natural constitution by which such a sensation gives rise to a conception of motion. There is here no convention involved at all nor any rule following in the sense of understanding a directive and intending to follow the directive. There is simply an operation of the faculty of perception, a computation in modern terminology, that takes us from the input of sensation to the output of conception. This happens without our having any intention that it should, and, indeed, we are, in fact, powerless to prevent it from happening. However, Reid also notes that such principles that operate quite automatically to yield conceptions may also be formulated and become principles of reasoning which are followed intentionally. Moreover, the decision to allow a certain word to express a concept, the word thus belonging to some conventional or artificial language that is the creation of mankind, is the decision to accept some social rule or convention. This decision presupposes conceptual understanding, however, and therefore, such decisions and intentions cannot be the basis for explaining our understanding of all empirical conceptions. Some empirical conceptions must originate in us by virtue of innate principles of our natural constitution that operate to produce those conceptions in response to natural sensory signs without intending they do so. This theory strikes me as an alternative or possibly a supplement to the one that Professor Körner has articulated. I should like to invite Professor Körner to explain to us how on his theory the regress and related problems may be avoided.

I know that the primary interest of Professor Körner in his book was to develop a logic of concepts, which he did with great sophistication, rather than to explore theories of the acquisition of concepts. The problems is to explain how his theory of concepts is consistent with the acquisition of concepts and whether it logically commits us to a theory of innate conceptual mechanisms like Reid's. If Professor Körner's theory does not commit him to a theory like Reid's, I am sure he will have an illuminating explanation of how the regress and circularity arguments may be avoided.

* * * * * * * * * *

REFERENCES AND NOTES

J.A.Fodor, **The Language of Thought**, (Cambridge, Mass., Harvard University Press, 1979).

S.Körner, **Conceptual Thinking** (Cambridge, University of Bristol and Cambridge University Press, 1955).

T.Reid, **Inquiry and Essays**, (eds.) K.Lehrer & R.Beanblossom (Indianapolis, Bobbs-Merrill Co., 1975). References to Reid's **Inquiry into the Human Mind on the Principles of Common Sense** are abbreviated to **CS** and to his **Essays on the Intellectual Powers** to IP.

W.Sellars, "Some Reflections on Language Games," in **Science, Perception and Reality** (New York, The Humanities Press, 1963), 321-358.

The research for this paper was supported by a grant from the National Science Foundation and was influenced by an earlier paper written by Lehrer in collaboration with J.C.Smith VI entitled "Reid on Testimony and Perception", **Canadian Journal of Philosophy** Supp.Vol.11 (1985). The author is grateful to J.C.Smith VI for his comments on an earlier draft of the present paper.

* * * * * * * * * *

REPLY TO PROFESSOR LEHRER

Stephan Körner

I am grateful to Professor Lehrer for discussing my early views on ostensive concepts, ostensive rules and their relations to each other and for raising some interesting points. They will allow me to explain and defend some aspects of my original position on the logic of inexact concepts and propositions as well as to say a litle about the way in which this position developed in the subsequent years. He rightly distinguishes between a logic of concepts and a theory of their acquisition and rightly says that my primary interest, as expressed in **Conceptual Thinking,** was the former. His objections are directed against what he takes to be the topic of my secondary interest, namely the acquisition of concepts. Yet it was never my intention to develop such a theory - though I obviously must have given this impression to at least one fairminded and highly competent reader.

I hope that he will be satisfied if I divide my answer into three sections. In the first I shall try to defend my early account of ostensive concepts against the charge that it involves a vicious, infinite regress and shall argue that ostensive concepts constitute an important species of inexact concepts, i.e. concepts admitting border-line cases. Next I shall in support of this defence very briefly outline my present views about the nature and function of inexact concepts in commonsense, mathematical and scientific thinking. I shall conclude with a few words about what I take to be the relation between on the one hand the logic of inexact and exact concepts and on the other hand various theories of the acquisition of concepts, e.g. by means of innate mechanisms.

1. Ostensive Concepts and Ostensive Rules

It seems best to begin by pointing out that I do not distinguish between accepting a concept and accepting the rules governing its use. To accept the concept 'table' (or one of the predicates expressing it) is to accept certain rules of reference according to which, e.g., a particular elephant is incorrectly asserted to be a table; and certain rules of inference according to which, e.g. 'being a table' is correctly asserted to imply 'being a piece of furniture'. This manner of speaking is by no means idiosyncratic. It is for instance, found in Carnap's writings, where accepting the syntax and semantics of a language is identified with the acceptance of certain rules of formation, transformation and satisfaction. It is also found much earlier in Kant's **Critique of Pure Reason** where the understanding is interchangeably characterized as the faculty of rules and the faculty of concepts. The acceptance of a concept or, what comes to the same, of the rules governing its use may be more or less explicit. To make them more explicit than they normally are is a logical task, while to describe their acquisition is a task of empirical or of "speculative" psychology.

In one of the passages, quoted by Lehrer, I say that "all conceptual thinkers accept ostensive rules and thus ostensive concepts". I could have also said, or added, that all conceptual thinkers accept ostensive concepts and thus ostensive rules. Both statements are incompatible with what Lehrer takes to be "the basic idea" of my account of ostensive concepts, namely that to "understand an ostensive concept one must accept an ostensive rule" (together with other rules, as has been pointed out earlier). For on my account - at least as I intended it - to understand an ostensive concept is to understand the ostensive and other rules governing it. It is **not** to accept these rules, but to know what it would be like to accept them. I may for example, understand the ostensive and other rules governing the concept 'x is a witch' (the ostensive rules referring e.g. to certain bodily features), i.e. know what it would be like to accept these rules, and yet refuse to accept these rules for the simple reason that I consider the concept empty.

Although the formulation of the rules governing an ostensive concept involves examples and counterexamples of its applicability, it may be very difficult to convey to somebody else which ostensive concept one has in mind. One of the difficulties pointed out by Lehrer, consists in conveying the respect or the respects in which the examples and counterexamples are meant to resemble each other. Yet this difficulty is of no importance if one is not - as I was not - concerned with the problem of how ostensive concepts, i.e. the rules governing their use, are conveyed, learned from others, taught to others or acquired, but merely with the logic of ostensive concepts and its difference from the logic of exact concepts. If one's purpose is a logical inquiry, then everybody may, in the words quoted by Lehrer from **Conceptual Thinking** "limit himself to the consideration of ostensive rules formulated by himself for himself".

Before turning to the charge that my account of ostensive concepts, i.e. the rules governing their use involves me in an infinite regress, it seems proper to admit that in my first book I should have said more emphatically that the use of ostensive concepts is not governed solely by ostensive rules, but that the set of rules governing these concepts must include such rules. Once it is agreed that to accept (reject, understand, consider etc.) an ostensive concept is the same as to accept (reject, understand, consider, etc.) the rules, especially the ostensive rules, governing its use, the charge of a vicious regress is easily shown to be groundless. The attempted demonstration of the vicious regress begins with the thesis: "Learning to use a language is learning to obey the rules of the language" which, if applied, to ostensive concepts and ostensive rules becomes "Learning to use an ostensive concept is learning to obey the ostensive and other rules governing it." Since for me "using an ostensive concept P" and "obeying the rules governing an ostensive concept P" are synonymous or, to be careful, intersubstitutable in all contexts and for all purposes which are of interest to me, the quoted sentence merely shows how one and the same situation or one and the same action can be described by two interchangeable locutions. The quoted sentence does not describe two different steps and thus does not even involve a harmless finite regress. It involves no more a regress than do the following two sentences: "Learning to be an oculist is learning to obey the rules governing the exercise of the oculist's profession" or "Learning to be an oculist is learning to be an eye-doctor". None of the above sentences claims to explain in any way what it is to learn a language, the use of the concept of being an eye-doctor or anything else. For one may well be satisfied with being able to follow certain rules without being able to explain how this ability was learned or otherwise acquired.

Ostensive concepts are a species of inexact concepts, i.e. of concepts which have positive instances, negative instances and border-line or neutral instances, which can with equal correctness be judged positive or negative (though not both, positive and negative). A correct statement to the effect that an object is a positive, negative or neutral instance of a concept expresses respectively a true, false or neutral proposition. As I later discovered, the syntax of true, false and neutral propositions and, generally, of exact and inexact predicates was fully worked out by Kleene. However, his interpretation of the truth-value of neutrality differs from mine, since on his view a proposition possessing the third truth value may eventually **turn out** to be true or false - an eventuality which excludes a free choice between making it true or false.[1] J.P.Cleave has shown that the three-valued logic of inexact concepts is related to Quasi-Boolean algebra, as the classical two-valued logic is to Boolean and the intuitionist is to pseudo-Boolean.[2]

Having clarified the logical status of ostensive concepts and of inexact concepts generally, a few words must be said about the special philosophical importance of ostensive concepts. It lies first of all in the fact that every perceptual predicate either is or entails an ostensive predicate, the range of whose applicability is demarcated by perceptually given examples and counter-examples. A second, philosophically important features of ostensive concepts is that they enable us to exhibit the structure of perceptual continuity, e.g. the relative location of perceivable and imaginable colours in a continuum of colour the parts of which are either subcontinua or common border-line cases of predicates characterizing subcontinua and of the complements of these predicates. In other words the logic of ostensive propositions and predicates enables us to make clearer the analysis of empirical - as opposed to Cantorian or Dedekindian - continua, undertaken by Aristotle and by Brentano.[2]

2. On Inexact Concepts and the Relation Between Commonsense and Theory

The further development of my views on the logic of ostensive and, more generally, inexact concepts was connected with my interest in the structure and application of mathematical theories and of scientific theories embedded in mathematics. In discussing these issues I still may occasionally have used locutions which are appropriate in discussions of the acquisition of theories, but I doubt whether anybody would regard such metaphors as pointing to an attempt at developing an empirically or speculatively psychological theory of the acquisition of mathematical or scientific concepts. A very brief indication of the relevance of the logic of inexact concepts to mathematical and scientific thinking might take the following form: Commonsense language, which the mathematical or scientific expert shares with semi-experts and laymen, differs from theoretical language both, in its logical structure and in its conceptual net. As regards the former difference, commonsense thinking is embedded in an inexact, theoretical thinking in an exact logic. In order to exhibit the way in which a scientific theory can be applied to phenomena described in commonsense language, one must draw attention to the following steps: description of empirical phenomena in commonsense language; idealization, i.e. transposition from commonsense language into the language of the theory; theoretical inference leading to a conclusion in the language of the theory; "de-idealization" of this conclusion into commonsense language; statement of the connection between the original and the final commonsense statement.

It is important to note that the structures described in commonsense language (however acquired) and the structures described in the exact theoretical language are not identical or isomorphic. Yet they are, in the context and for the purpose of the theory's use, treated **as-if** they were identical. As has been pointed out by some of my critics, and by myself, my account of the application of mathematics and mathematized theories in terms of idealization, de-idealization and as-if identification recalls Plato's theory of Forms and of the methexis of the perceptual world in them. There are, however, important differences - especially the difference between Plato's realist assumption of one and only one realm of Forms and my assumption of the possibility of alternative Platonic heavens, e.g. Euclidean and non-Euclidean ones. I may add that in opposition to followers of Frege who believe that commonsense language needs to be purified by removing its inexactness and in opposition to followers of Wittgenstein who regard such "purification" as totally misleading, I try to explain that and why both a "Fregean" exact and a "Wittgensteinian" inexact language are needed.[3]

3. On the Relation between the Logic of Exact and Inexact Concepts and the Acquisition of these Concepts

So far I have tried to defend myself against the charge that my original account of ostensive concepts and its later development imply a theory of the acquisition of concepts which involves a vicious infinite regress. The defence was based on the argument that since I have no theory of the acquisition of concepts, there is no case to answer. If I have given the impression of trying to develop such a theory, I should plead the extenuating circumstances of an author writing his first book in a language which he is still trying to acquire.

It seems clear that human beings, unlike lower animal species, have the native ability to acquire concepts. Philosophers have therefore either spoken of man's possession of innate concepts of which he had the ability to become explicitly aware in the course of his life or of his disposition to acquire such concepts. Lehrer has some sympathy with a theory of the first kind. He agrees with Thomas Reid's thesis that "if mankind had not an (original) language" i.e. an original way of using concepts or conforming to rules for their use, "they could never have invented one by their reason and ingenuity". He further suggests that by accepting Reid's or a similar theory I might be able to produce "a consistent modification" of my theory "rather than a systematic alternative to it". To this characteristically fair suggestion, my answer is as follows: Even though proposing a logical theory of conceptual thinking and proposing a theory of the acquisition of concepts are separate undertakings their results may be mutually incompatible - e.g. if the genetic theory assumes that conceptual thinking possesses features the presence of which the logical theory denies. My own logical theory - in its early phases and, more obviously, in its present form - **is** compatible with Reid's theory. But it may also be compatible with other theories, which are unknown to me and which in addition to the aesthetic appeal of Reid's theory, can claim more empirical support.

* * * * * * * * * *

FOOTNOTES

1. See S.C. Kleene, **Introdution to Metamathematics** (Amsterdam, 1952) Section 64.

2. See J.P.Cleave, "Quasi-Boolean Algebras, Continuity and Three-valued Logic", **Zeitschr. für Logik und Grundlagen der Mathematik** 22 (1976) pp.481-500.

3. For the indispensability of commonsense language in legal thinking see my "Sprachspiele und rechtliche Institutionen", **Proceedings of the 5th Wittgenstein Symposium** (Wien, 1981) pp.480-491.

* * * * * * * * * *

3. FIVE CONCEPTS OF FREEDOM IN KANT

To Stephan Körner on our Seventieth birthday,
September 26, 1983

Lewis White Beck

In Kant's works I can distinguish at least five important conceptions of freedom. In part they overlap, some are inconsistent with others, and some presuppose others. Kant's nomenclature for them is variable, and for some of them he has no name at all. Three of them appear to me to be untenable, and the others which are more promising are hardly more than merely adumbrated by Kant. In **The Actor and the Spectator** I gave a fuller development of these latter conceptions, and though I briefly indicated their Kantian character I did not show their Kantian provenance. After briefly presenting the better-known Kantian conceptions of freedom, I shall devote the remainder of the paper to documenting, elaborating, and defending the fifth conception, and showing its relation to Stephan Körner's.

1. The Empirical Concept of Freedom

The first concept of freedom is empirical, since empirical considerations are decisive for the question whether a specific act was freely done. The kind of freedom involved is called by Kant the comparative, the psychological, and the practical.[1]

It is an empirical question whether, and to what extent, a specific person in a particular case acted freely. "Freely" here means voluntarily, not coerced. Aristotle[2] formulated criteria of freedom in this sense, and today there are specific juridical criteria of it. In deciding whether to impute responsibility to a lawbreaker, one raises empirical, not metaphysical, questions. We want to know, for instance, whether he was driven by an uncontrollable passion. When we decide that a particular person was free in doing a particular action, we say nothing about men in general and freedom in general. A man may be free today, but not free tomorrow. It is not this concept of freedom, but solely its moral sufficiency[3] which Kant later rejected. In the **Critique of Practical Reason** he calls it, as a foundation of morality, a "wretched subterfuge" (**elender Behelf**) which grounds only the "freedom of a turnspit."[4]

2. The Concept of Moral Freedom

It is the task of the **Critique of Practical Reason** to confirm freedom in the transcendental sense. But since I have not yet discussed this concept of freedom, I must say here in only a preliminary way: the concept of freedom in the second **Critique** is the concept of a freedom which cannot be empirically established and which is not, like the freedom Kant calls comparative, a peculiar kind of natural causality. The **Critique of Pure Reason** had proved, Kant thought, the possibility of transcendental freedom; in the second **Critique** he attempted to show its actuality. The freedom whose actuality is necessary for morality is transcendental freedom. But the ultimate identification of moral with transcendental freedom is not self-evident on the surface; in themselves they appear to be independent of each other. Only after I have developed the concept of moral freedom will I come to the concept of transcendental freedom in its own right and show the relationship between them.

The main steps in the proof of moral freedom are as follows. Granted the moral phenomenon, "the fact of reason"; i.e. assume that the ground of moral action is a pure law. If this law and the decision to act out of respect for it were links in an empirical temporal chain of natural causes, the moral phenomenon as described would be impossible. But moral action as described is not impossible; therefore moral action cannot be an effect of natural causes.[5]

The moral law is the **ratio cognoscendi** of freedom, and freedom is the **ratio essendi** of moral responsibility.[6] A free will and an unconditionally commanding practical law presuppose each other.[7] The mere independence of the will from the incentives of sense and feeling is freedom in the negative sense, and it, like empirical freedom, is a precondition of freedom in the positive sense, namely the effectiveness of the legislation of pure practical reason and the ability to undertake actions in accordance with and because of (out of respect for) this law. Freedom in this positive sense is called **autonomy**.[8] The situation, however, is more complicated than it appears in the second **Critique**. In the **Metaphysic of Ethics** Kant distinguished between **Wille** and **Willkür**. **Wille** is pure practical reason which, through its autonomous legislation, creates moral duty in a being who does not by nature adhere to the law. It is by **Willkür** that an action in accordance with, or opposed to, this law is undertaken. If one continues the political metaphor of autonomy, one can say that **Wille** is the autonomous legislative function and **Willkür** exercises the executive function. The freedoms of the two are therefore conceptually different. Kant even says: "**Wille**, which is concerned with nothing else than law, can not be called either free or not free, because it is not concerned with actions but directly concerned only with legislation for maxims... Only **Willkür** can be called free."[9] And in the **Critique of Judgement** he says: "Where the moral law speaks, objectively there is no further free choice with respect to what is to be done."[10] Consequently one must distinguish as follows:

(a) The moral autonomy of **Wille**, i.e., its independent authorship of law independent of motives and impulses of the empirical, sensible world; and

(b) The moral freedom of **Willkür**, i.e., its capacity of spontaneously obeying the law as its maxim and thereby inserting a new link in the causal chain of events in the sensible world.

Before Kant explicitly drew this distinction, autonomy and feedom and **Wille** and **Willkür** were almost interchangeable terms, and he spoke not only of the freedom of **Wille** but even of the autonomy of **Willkür**.[11] In spite of these confusions I will speak of moral freedom simpliciter when I refer to the concept of freedom which is analytically connected with the concept of the pure moral law. It is the kind of freedom manifest in and only in genuine moral action - not even in merely legal, unmoral, or immoral actions.[12]

3. The Concept of Freedom as Spontaneity

"Freedom as Spontaneity" is not entirely suitable as a title for this conception, since Kant uses "spontaneity" with respect to both moral and transcendental freedom, and he does not give a specific title to the broader, nonmoral conception to be described in this section. I can, however, think of no other title which might not be as misleading as this.

The argument for moral freedom had as premise the consciousness of the moral law and, more specifically, the moral law in its specifically Kantian formulation. A free action and an action done out of respect for the law are the same. Here arises an aporia. An unmoral (legal) or immoral action is not done because of the law (even if, as legal, it conforms to the law), but is done on account of subjective, individual, empirical impulses. A nonmoral or an immoral act is to be entered to the credit of the mechanism of nature and is therefore not morally imputable. But to do justice to the ethical phenomenon, one must have a concept of freedom which permits the imputation of unmoral and immoral actions. The criteriological or analytical connection between freedom and morality (positive moral value) must be loosened.

Accordingly, in **Religion within the Limits of Reason Alone** Kant argued as follows. Since imputable but immoral (evil) actions occur, **Willkür** must choose its guiding maxims freely and not from the necessity of nature. The "act of **Willkür**" must have two functions:

(a) "the use of freedom whereby the highest maxim (be it in accordance with, or opposed to, the law) is taken up into **Willkür**"; and

(b) the use of freedom "such that the action itself... is exercised according to that (freely chosen) maxim."[13]

Even if the maxim is opposed to the law, the action done under it can be free and imputable. This kind of freedom, unnamed by Kant, is unlike empirical freedom, because it is inexplicable by natural causality; and it is unlike moral freedom, because it is not a **sufficient** condition of positive moral worth. In fact, Kant does not use this concept of freedom exclusively in his treatment of morality at all. He does not systematically discuss the concept in this, its broader, use, but I find two passages which concern this broader use.

The first passage is the familiar one at the beginning of Part III of the **Foundations of the Metaphysics of Morals,** and its position in that work may suggest that Kant is thinking about freedom in a moral context. But the argument is not limited to the ethical alone:

> "Now I say that every rational being which cannot act otherwise than under the idea of freedom is thereby really free in a practical respect... Now I affirm that we must necessarily grant that every rational being who has a will also has the idea of freedom and that it acts only under this idea... Now we cannot conceive of a reason which consciously responds to a bidding from the outside with respect to its judgments, for then the subject would attribute the determination of its power of judgment not to reason but to an impulse. Reason must regard itself as the author of its principles, independently of foreign influences."[14]

The remainder of the paragraph, in speaking of practical reason, does integrate this consideration with moral philosophy; but enough has been said already to indicate that, even in thinking, reason must regard itself as autonomous and not as mechanically responding to natural causes. This is the explicit theme of the following passage: The philosopher who denies freedom "has deeply in his soul, although he does not want to confess it, presupposed that the understanding has the faculty of determining his judgment according to objective grounds which are always valid, and that it does not stand under the mechanism of merely subjective determining causes which can vary in their consequences."[15] In this confrontation with the denier of freedom, Kant does not begin with the moral consciousness (though in earlier parts of his review he has repeated the more usual moral argument), which might conceivably be explained away on psychological grounds. Rather, he begins with the phenomenon of decision in general. One could almost speak of autonomy in a non-moral sense. Against Schulz, who denied freedom, Kant continued: "He always assumed freedom to think, without which there is no reason" - and this includes theoretical reason and our whole cognitive capacity, for thinking is also a kind of deciding and acting. Everyone who decides presupposes that his decision was not causally determined in advance.

After I have reached a decision, someone can explain to me the causal grounds on which he had already in advance foreseen what my decision was going to be, for my character traits and the relevant psychological and physiological laws made my decision and action causally necessary. Indeed I myself can subsequently sometimes explain my own actions in the same way; I regularly do so when I excuse myself, for, as Sartre says, "Psychological determinism is an attitude of excuse."[16] I say that I was not responsible, for my behavior had causes which did not lie within my power, which were at the time unknown to and therefore uncontrollable by me, and which reduced my "decision" to an illusory experience (like a "decision" made as a consequence of hypnotic suggestion). **Post facto** I can be a fatalist about what I have done; but if I did not in the moment of choosing have this experience of free choice, the phenomenon of choice would not merely be illusory - it would not even exist as an illusion. I experience no choice or decision (that is, I do not choose or decide) without believing that my decision or choice will make a difference, that is, without believing that the outcome is not already causally fixed. My experience of freedom of choice - that the choice is "up to me" and makes a difference in the outcome - is **schlichtgegeben** and has a self-evidence greater than any possible causal law which would entail its necessary illusoriness.

I am not asserting that this phenomenon is **in every case** self-certifying; we do have the counter-example of post-hypnotic "decisions."[17] The mere experience of decision does not prove the reality of freedom, for it is a genuine experience of decision only when the decision itself is really free. A madman believes he decides something (i.e., he has the experience of deciding) when in fact he doesn't possess even **empirical** freedom. Mephistopheles says: "You believe you push, but you are pushed." Genuinely deciding means, analytically, freely deciding; but the question is still open whether genuine decisions do take place. According to Kant moral freedom, and presumably the freedom which has been analyzed in the present section of this paper, would be impossible, in spite of their **Schlichtgegebenheit**, if there were no freedom in the transcendental sense.[18]

4. The Concept of Transcendental Freedom

The concept of freedom discussed in the foregoing section might be called the **transcendental concept** of freedom, a "transcendental concept" being Kant's name for one which underlies the possibility of a priori cognition; for without that concept of freedom, Kant tells us that reason (both theoretical and practical) would be impossible. The concept we are now about to discuss is the one Kant calls the **concept of transcendental freedom** meaning, I think, a transcendental concept of a peculiar kind of freedom which in all strictness might better be called **transcendent freedom** because it deals with a matter which transcends the limits of possible experience and the knowledge of theoretical reason. But the name "transcendental freedom" is too deeply entrenched in Kant's writings to make such a revision feasible.

In his own words, which in this instance are not highly accurate, the **Critique of Pure Reason** attempted to show not the actuality but only the possibility of freedom.[19] If freedom and the causality of nature were contradictory, there could be no freedom (except in the empirical sense).[20] It was the task of the first **Critique** to show that they did not contradict each other; it was reserved for the second **Critique** to show that the freedom shown possible in the first (transcendental freedom) is actual (moral freedom). So much for the formalities of the argument; in fact some of the first **Critique** is concerned with the reality of freedom, and some of the second with its possibility.

Kant's theory of transcendental freedom is so well known that it can be recapitulated with great brevity. On purely epistemological grounds Kant showed that nature is a unitary causal system in which each state or event has a sufficient condition in preceding and simultaneous states or events. The thesis of the Third Antinomy proved that natural causality is not the only causality, and that there is also a non-temporal cause, i.e., that there is a causality of a free cause which is not causally determined by antecedent or contemporaneous events. The proof of the thesis is hardly more than a repetition of the Aristotelian-Thomistic proof of the reality of a primordial cause of the world, which is not a consequent of any antecedent condition and therefore not causally determined. This thesis is opposed to the antithesis, which proves the universal validity of natural causality and excludes free causation from the course of nature and also a first cause.

Kant believed that he proved both. The solution of the problem arising from proofs of the truth of two propositions which contradict each other he found in his famous "two-world theory." According to this, there is a phenomenal world, wherein each change is determined by an earlier one in space and time; and a noumenal world, which is not spatial and temporal, and of which the phenomenal world is only an appearance to minds constituted like ours. Free causality within the noumenal world and between the noumenal and the phenomenal can be **thought** without contradiction, but only the temporal causality relating events and states in the phenomenal world can be **known.** There is no contradiction, because free causation and natural causation are predicated of ontologically different kinds of beings. Consequently the causality of a thing in itself in the noumenal world can be thought of as free, while its appearance in the phenomenal world can be known (in principle) without exception as causally necessitated. One and the same action is to be regarded as free, inasmuch as it arises in the reality of the noumenal world, and as causally necessitated, since it occurs in the complex of phenomena of the spatio-temporal world of appearances.

The thesis of the Third Antinomy is explicitly concerned only with the question as to whether there was a (free) creation of the world. In showing (to his own satisfaction) that the thesis can be true without conflicting with the "reign of law" in nature, Kant says that he has only shown the **possibility** of freedom, and has given a sufficient answer to those who would argue against freedom on the ground that it is impossible because inconsistent with the mechanism of nature. If a philosopher finds good grounds for asserting the freedom of human action, the fatalist can no longer object that freedom is impossible because it would overthrow natural causality. But (except hardly as **obiter dicta**) the first **Critique** does not argue for the reality of transcendental freedom.

The **Critique of Practical Reason** finds in morality good grounds for the assertion of transcendental freedom by showing that moral freedom presupposes it (or is identical with it). But this redemption of freedom from the universal mechanism of nature proves both too much and too little. Too much: for according to it every phenomenon has its transcendental ground or noumenal causation. A proof which shows that if anything is free then everything is free proves too much. One wants to show that only the empirical, moral, or spontaneously free action is transcendentally free, and not that just any event, a bodily reflex or an apple falling off a tree, is free.

The theory proves also too little. Kant himself concedes that according to it, judgments "at first appear to conflict with all equity."[21] But is this only **prima facie**? Why should one rue an unjust deed, when the deed was unavoidable? How can one maintain that a man is responsible for his actions, and at the same time assert: "Every one of the voluntary actions is determined in advance in the empirical character, before it ever occurred"?[22] May one assert that a man is free after one has declared that if we knew all the empirical facts and the natural laws of their connection, we would be able "to compute his behavior with certainty, as we do an eclipse of the sun or moon"?[23]

If transcendental freedom means noumenal causality, the ubiquity of noumenal causality trivializes the concept of freedom. Since the noumena and their causality are unknowable, there is no possibility, in research into human phenomena, to determine why in some individual cases we should be allowed to apply the concept of moral freedom and in other cases forbidden to do so. The uniformity of human actions - including moral actions - is in principle as great as that of the solar system, according to Kant's theory of nature. Granted this and the ubiquity of noumenal causation there is no reason why assertions of the freedom of our actions should have any consequences different from denials of that freedom. (There will be consequences either way, which seem to "conflict with equity". We assume the freedom of the noumenal man, but we hang the phenomenal man.) If the faculty of transcendental freedom has any meaning for the course of nature and history - i.e., if free men act differently from unfree men by virtue of their freedom and not by virtue of different natural causes (education, environment, heredity, and the like) - then the uniformity of nature is abrogated. And if it has no consequences for the uniformity of nature, moral freedom, which Kant holds depends upon it, is an empty pretension.

5. The Concept of Postulated Freedom

To remove, or at least to ameliorate, these aporias we must investigate a passage in the first and one in the third **Critique**, where we find a sketch of a theory of freedom as a postulate or as a heuristic maxim, and revise the common interpretation of Kant's ontology as a "two-world theory."

When I speak here of freedom as a postulate, I am not referring to the "postulate theory" of the **Critique of Practical Reason.** According to it, a postulate of pure practical reason is an assumption inextricably connected with morality yet theoretically unprovable. In this sense Kant held the immortality of the soul and the existence of God to be postulates and in one place he counts the freedom of the will among the postulates.[24] There is (with respect to freedom) nothing new in this; it only makes explicit what had already been indicated in discussion of the concepts of freedom in the Analytic of Pure Practical Reason, namely, that one cannot act without presupposing a theoretically unprovable freedom.

Instead of to this well-known conception of postulates, I am referring here to a conception which Kant did not work out in full, and in the discussion of which he does not even use the word "postulate". In the **Critique of Pure Reason** Kant makes a remark which suggests not only the other theory of postulates but also a revision of the two-world theory. He says:

"...As regards (man's) empirical character, there is no freedom,

> and yet it is only in the light of this character that man can be studied - if, that is to say, we are simply **observing,** and in the manner of anthropology seeking to institute a physiological investigation into the motive causes of his actions. But when we consider these actions in their relations to reason - I do not mean speculative (i.e., theoretical) reason, by which we endeavour **to explain** their coming into being, but reason in so far as it is itself the cause **producing** them - if, that is to say, we compare them with (the standards of) reason in its **practical** bearing, we find a rule and order altogether different from the order of nature."[25]

Instead of thinking of two worlds, one noumenal and one phenomenal, Kant is here thinking of one world under two aspects. There is not one **homo noumenon,** who is free, and a **homo phaenomenon** who is not. The noumenal and the phenomenal are not ontologically distinct (like an object and a picture of it) but are aspects determined by methodological procedures chosen with regard to the divergent purposes of two kinds of inquiry.[26] Each of these aspects is determined by what I shall call its respective postulate. The postulate theory of freedom, and the double-aspect theory of noumenon and phenomenon, are two expressions of the same fundamental theory. From the scientific point of view of the observer, causal conditions are sought and (in principle) found. From the practical point of view of the acting man (or the man who is normatively judging another's actions), the reasons for decisions and actions are sought and evaluated. It is only from the second standpoint that an action can be interpreted as free and imputable.[27]

It is not until the **Critique of Judgment** that Kant gives much help in developing this postulate theory of freedom. There he presents an antimony between teleological and mechanical explanation which readily suggests by analogy the theory of freedom and natural cause as postulates. The Antinomy of Teleological Judgment is analogous to the Third Antinomy of the first **Critique,** but its solution is wholly different. The analogy is between natural law and moral law (in the first **Critique)** and mechanical law and teleological law (in the third). The solution of the Antinomy in the first **Critique** is, as we have seen, ontological (the two-world theory); the solution of the Antinomy in the third **Critique** is purely methodological and postulational. In consequence, the Second Analogy of Experience in the first **Critique,** which gives rise to the antithesis of the Third Antinomy, is interpreted by Kant himself, in the third **Critique,** as a methodological maxim or postulate. The effect of this reinterpretation upon the ontology of the first **Critique** is a profound one. It is no longer the case that one of the laws (the law of natural causality) is given a pejorative standing as a law of appearance only of what is ultimately real and true. This is evident in the third **Critique**'s reformulation of the Second Analogy, which is now read as a methodological rule: "All production of material things and their forms must be held to be possible according to purely mechanical laws." As a maxim it does not contradict the other maxim of judgment: in the explanation of organic forms seek the purpose of each member.[28] Nor does it contradict the postulate underlying action and the judgment of actions, which is that actions must be imputed to free agents.

Had Kant explained in his solution of the Third Antinomy that the antithesis is not a constitutive principle of nature but only a regulative idea or methodological maxim - a conclusion he reached only in 1790[29] - he would not have found it necessary to defend two opposed philosophical theorems by the desperate ontology of the two-world theory. He could have lived happily with two postulates which do not conflict because they are not used at the same time or for the same purpose. They are:

(a) **Postulate for the scientific explanation of human actions**: In natural sciences always seek natural causes, and do not admit non-natural causes in the explanation of natural phenomena (including human actions).

(b) **Postulate for ethical and practical decisions**: Act as if the maxim of your will were a sufficient determining ground for the action undertaken.

(Corollary: **Postulate for the normative evaluation of another's action**: Judge as if the maxim of the will were a sufficient determining ground for the action in question.)

None of these postulates makes an ontological claim. Rather, they tell us what we must do in order to execute the function of a scientific observer, an agent, or a judge.

In spite of the words "as if" in the formula of the second postulate and its corollary, the theory I am proposing here is not a form of a dogmatic fictionalism which asserts that men are not free, but that in practice one must think and act as if they were free. Such fictionalism presupposes mechanical determinism as a metaphysical truth, in comparison with which freedom is **only** a fiction. The view I find in Kant and am here defending holds that there is as much - and as little - ground to hold **all** the postulates as fictions. They are on an equal footing and, in different spheres of experience, equally inescapable. Their truth is not absolute, for they limit each other; their truth is in the context of their respective employment definite and justified; outside the respective contexts there are no criteria for their correct application.

In this conception we are no longer, as in orthodox Kantianism, forced to assign to the sciences the realm of the appearances (in an ontologically pejorative sense) and to ethics the realm of noumena (in an epistemologically pejorative sense.) The **a priori** structures of each context remain intact, and each claims to govern the entire relevant experience of its realm.

When we accept this revision of Kantian transcendental theory, we may regard the realm of practice as one aspect of the scope of our experience in general, which, under other postulates and concepts, appears as the realm of nature. It is almost the same to hold the causal concept of the Second Analogy as regulative for the investigation of nature, and to hold the practical concept of freedom to be constitutive for the realm of human actions, as it is to hold with Kant in the Methodology of the first **Critique** that natural causality is a constitutive concept and free causality a regulative Idea. The more mature theory, towards which I think he was moving even before the **Critique of Judgment**, holds that each is regulative, or each is constitutive of its own realm. According to Kant we know that a change is an objective event only under the condition that we know it to have a cause; analogously, we know that a human event is an action only under the condition that we know it to have reasons and not merely causes. Each of these definitional conditions depends upon the methodological postulates in use.

6. The Complementarity of Freedom and Natural Necessity

There is an analogy between the theory outlined here and the principle of complementarity in physics. In physics one and the same thing must be described sometimes as a wave and sometimes as a particle. Which description fits is dependent upon the conditions of its observation, which cannot be simultaneously fulfilled. The question of what the thing itself is, irrespective of the conditions of its observation, is unanswerable; but it is a question that need never arise. Similarly here. One cannot give a context-free answer to the question: Is this man, or this piece of behavior, free? Nor can one, like Kant, ascribe both freedom and natural necessity to the same action. **But one does not ever need to do so, either for the sake of science or of morals.** The remainder of this paper will discuss the reasons for this.

Looked at abstractly and schematically, freedom and natural necessity appear to exclude one another. Accordingly, Kant's version of compatibilism, his two-world theory, has the appearance of a paradox, since it asserts both at the same time of the same action. Just so, "the wave theory of the electron" and "the particle theory of the electron", taken abstractly and schematically, appear to contradict each other.

Looked at concretely and empirically, however, no occasion for contradiction arises in either case. In practice, one never has occasion to say of an electron that it is acting like a wave and acting like a particle at the same time in the same observational field; nor does one ever have occasion in a situation calling for decision or evaluation to affirm both the thesis and the antithesis of the Third Antinomy or of the Antinomy of Teleological Judgment. In practice, one chooses between them. In no single case does one say: "The criminal behavior of the defendant was caused by a specific brain injury (or a specific physiological state, or the like) and therefore was as unavoidable as an eclipse of the sun; **nevertheless** the defendant acted freely and is therefore responsible for his action." If I did not, as a Kantian philosopher, hold **a priori** that any action whatsoever was causally determined, but rather as a forensic expert concretely and empirically did know precisely what empirical cause produced the action under investigation, then I would no longer say (with Kant) that this action was a free one and the defendant was responsible for it. In a court of law opposing counsel dispute whether the behavior in question was necessitated (or at least made highly probable) by definite, precise, perhaps pathological conditions; and for this they have criteria of empirical freedom more sophisticated than those of Aristotle and common sense. But whether the causally necessitated action was **also** a free action - outside the philosophical seminar and lecture hall that is never the question.[30]

The theory of freedom as a postulate for practical and moral life and of natural necessity as postulate for research into nature and human nature enables us to draw from empirical observation the practical conclusion as to whether in a specific case a specific human being behaved freely or not. We have gained a position from which empirical freedom can be taken with conceptual seriousness. If we do not assume, merely schematically and **a priori**, that every action has a natural cause, but discover empirically under the governance of this postulate natural causes sufficient to explain it, the postulate of natural causality is in so far forth confirmed, and we rescind the imputation of freedom and responsibility. Since in most cases we are not capable of perfecting such causal explanations, the question often remains unanswered whether a specific man in a specific situation acted freely or not. But according to this theory that question is answerable in principle, whereas a practical, contingent answer could not be supported by a theory of transcendental freedom based upon the two-world theory ordinarily associated with Kant's name.

Kant said that the theory of transcendental freedom appeared **prima facie** to conflict with equity. It must be granted here, too, that if we act on the theory of freedom as a practical postulate we may sometimes unjustly impute to a man an action which he could not have left undone. In this case, we commit an error and do an injustice. It may also happen that we abstractly and schematically explain an action causally for which no **sufficient** causal explanation can in fact be given. In this case we do not accord the man the dignity which belongs to him. Since we cannot empirically determine the limits of our explanatory schemata, such scientific dogmatism, and such injustice, are perhaps unavoidable.

In both cases we judge unjustly, sometimes imputing unfree actions, sometimes denying responsibility where it actually exists. But the alternative to sometimes making such mistakes is - always making such mistakes. If we assume the two-world theory of transcendental freedom and phenomenal necessity, I think we **always** judge wrongly when we hold a man responsible for any of his actions, since, according to the first and second **Critiques,** his action could not have remained undone. I believe that if Kant had rewritten the **Critique of Practical Reason** in the light of the **Critique of (Teleological) Judgment** he would have been better able to avoid moral judgments "which appear to conflict with equity."

It is gratifying to find that, following a very different path, I have come to conclusions so consonant with those of Stephan Körner in his British Academy lecture, "Kant's Conception of Freedom" (1967). The alternative standpoints and postulates in my account correspond very closely to the alternative categorial schemata (p.195) and ideal concepts (p.212) in his revision of Kant. I hope that my formulation of the principal conclusion is as acceptable to him as his is to me; for, as he says, unless one denies the uniqueness of the Kantian categorial schema of nature, "effective freedom can only be saved in Kantian fashion by being located in the noumenal world" (P.210), but the rejection of its uniqueness "removes (freedom) from the noumenal world and places it firmly into nature": (p.214). I would venture to disagree only with the very last word, "nature". What I conceive of as nature is under categorial schemata which exclude freedom. I would substitute for "nature" his expression "world of experience" (p.211, last line). It is the world of (precategorial) experience which provides material for a diversity of categorial schemata, including those of nature, morality, and culture.

* * * * * * * * * *

FOOTNOTES

1. **Critique of Pure Reason** A802-B830; **Critique of Practical Reason**, Akademie ed., V, 96-97.

2. **Nicomachean Ethics**, Book III, Ch.1.

3. See **Critique of Pure Reason** A803-B831. On different interpretations of this passage, see my **Commentary on Kant's Critique of Practical Reason**, p.190, note 40.

4. **Critique of Practical Reason**, 96, 97.

5. **ibid**, Section 5.

6. **ibid**, p.4, note.

7. **ibid**, Sections 5,6.

8. **ibid**, Section 8.

9. **Metaphysics of Morals**, Ak.ed. VI, 226.

10. **Critique of Judgement**, Section 5 Ak.ed. V, 210.

11. **Critique of Practical Reason**, 36.

12. **ibid**, Section 8 end.

13. **Religion Within the Limits of Reason Alone**, Ak.ed. VI, 31.

14. **Foundations of the Metaphysics of Morals**, Ak.ed. IV, 448.

15. **Recension zu Schulz's Sittenlehre**, Ak.ed. VIII, 14. Körner, **Proceedings of the British Academy** LIII, 203 (1967), finds a like thought in Kant's Reflexion 4904.

16. **Being and Nothingness** (New York, 1956) p.40. Kant accepts no excuse (**Critique of Pure Reason** A555-B583; **Critique of Practical Reason**, 98). A more indulgent attitude to human weakness and the strength of nature is found in the note to **Critique of Pure Reason** A551-B579.

17. In **The Actor and the Spectator** I have argued that it is self-confirming only in the case of assent to an argument concerning the causation or reasons for this assent.

18. **Critique of Pure Reason** A534-B562; **Critique of Practical Reason**, passim.

19. Only the logical possibility, not the real possibility or the actuality, is established, according to **Critique of Pure Reason** A558-B586.

20. **Critique of Pure Reason** B xxix.

21. **Critique of Practical Reason**, 99.

22. **Critique of Pure Reason** A553-B581.

23. **Critique of Practical Reason**, 99.

24. **Critique of Practical Reason**, 132. The three postulates are not cognate. Those of immortality and the existence of God are necessary for the **summum** bonum, while that of freedom is necessary for morality itself. See my **Commentary**, Ch.14.

25. **Critique of Pure Reason** A550-B578.

26. This is not the place to reconstruct the theory of the thing in itself and the noumenon along the lines presupposed in the present discussion. Though Kant often writes almost as if the thing in itself were a Lockean substance (an "unknown cause of my sensations") and as if the phenomena and things in themselves were numerically different things of different ontological kinds, he also writes as if the thing perceived can be thought (not known) as existing in itself without regard to the forms under which it is known. Regarded in this way the very thing perceived is the thing in itself. We perceive the thing itself but not as it is **in** itself. As an object of thought but not of knowledge, it is properly called noumenon; as an object of perceptual knowledge, it is properly called phenomenon. The denial of the two-world theory in favour of a two-aspect theory (the phenomenonal and noumenal character of a single thing) was undertaken in my **Commentary on Kant's Critique**

of Practical Reason, pp.192ff., and has been ably supported by Gerold Prauss, **Erscheinung bei Kant** (Berlin, 1971) and **Kant und das Problem der Dinge an sich** (Bonn, 1974); and by Henry E. Allison "Things in Themselves, Noumena and the Transcendental Object," **Dialectica** 32 (1978) 41-76.

27. The previous footnote has pointed out the relation between thing in itself and noumenon. These terms may have the same denotation, but their connotations are quite different. If we do not identify them, and if we deny ontological dualism between either of them and a phenomenal object, and if we think with Kant of reason (as "the higher faculty of desire") as especially concerned with morals and of understanding as exclusively a cognitive faculty, then it makes perfectly good sense to say that a human being as an object of ethical judgment, and thus as a free agent, is a **homo noumenon**. This does not mean that he is a ghostly non-temporal, non-spatial unknowable **Ding an sich**. This accords with Kant's own usage in the **Rechtslehre** (Ak.ed. VI, 335). Thus, after all, it is the noumenal man who gets hanged.

28. **Critique of Judgment**, Sections 70-76.

29. Once (**Critique of Pure Reason** A798-B826) he called the antithesis of the Third Antinomy a maxim, but he did not exploit this meaning. The word "regulative" as applied to the Analogies of Experience is stated (A180-B223) **not** to have the significance which later the **Critique of Judgment** ascribed to it.

30. I do not speak contemptuously of the seminar room and the lecture hall, but the kinds of investigations which go on in them are not designed to discover actual causes of behavior, and my argument is that when we are convinced that we know the actual causes we must rescind the imputation of freedom, whereas in the philosophical lecture hall we can entertain a schematic, abstract, compatibilism, which Körner (**op.cit.**, 209) wittily compares to eating one's cake and having it too. In **The Actor and the Spectator** (pp.105-107) I pushed the analogy with the principle of complementarity to the point of showing that the conditions under which a specific, concrete causal explanation of an action is given **prevent its being** a free action, and not merely add another conceptual determination which **prevent its being interpreted as** a free action

* * * * * * * * * *

REPLY TO PROFESSOR BECK

Stephan Körner

Professor Beck's paper has - as was to be expected - proved of great exegetic and philosophical value to me. His clear distinction between five Kantian conceptions of freedom should advance any student's and scholar's understanding of Kant's doctrine of the relation between moral freedom and natural necessity. I, for one, certainly intend to use his discussion in any future course on Kant's philosophy. Beck expresses his discomfort about Kant's approach to the problem of the Third Antinomy and his solution of the problem by pointing out that, if one accepts that solution, then one must be content to "ascribe freedom only to the noumenal man", yet nevertheless "hang the phenomenal man". Beck is right in thinking that I share his discomfort and that my own view resembles the view which he proposes and defends in his book **The Actor and the Spectator**. Since in what follows I shall dwell on some differences between our views, it seems proper to emphasize at the outset that I regard our points of agreement as far more important than our differences. Indeed the differences between us have their source not in our approach to ethics but to the philosophy of science. In the following brief remarks I shall begin by trying to formulate the problem of effective freedom and various proposals for solving it, including the proposal on which Beck and I agree (Sect.1). I shall then give some of my reasons in support of this solution (Sect.2) and shall conclude with a brief comparison of Beck's account with my own. (Sect.3).

1. The Problem of Moral Freedom

In performing a voluntary action a person is capable of distinguishing between his chosen bodily conduct, i.e. a chosen bodily movement or non-movement; the course of nature which he believes to have preceded this movement, more briefly the retrospective course of nature or the retrospect; and the course of nature which he believes will follow his chosen bodily conduct, more briefly the prospective course of nature or prospect.

The person is, moreover, under the impression that (1) the retrospect and the bodily conduct together predetermine the prospect, that (2) the retrospect does **not** predetermine the bodily conduct and that (3) the retrospect does **not** (by itself, i.e. without the bodily conduct) predetermine the prospect. The action which he regards himself as performing is, furthermore, characterized by certain physical, cognitive and evaluative features which constitute its physical, cognitive and evaluative demarcation in the sense that the absence of any of these features would imply that he intended to perform a different action from that which he actually performed. Thus if the action considered is the paying of a debt, it is as a rule physically irrelevant whether the payment was done by cheque or by cash; cognitively irrelevant whether the debtor regarded his debt as avoidable; and evaluatively irrelevant whether or not the debtor disliked paying the debt.

For our present purpose we may, without loss, assume that our concept of pre-determination is a concept of causality and not e.g. one of increased probabilification. For the sake of brevity, we shall further make the restrictive assumption that the concept of effective freedom is to be considered only in relation to the world of experience or, in Kantian terminology, to the phenomenal and not the noumenal world, if any. This restriction saves us any detailed consideration of the Kantian doctrine that man is determined **qua** phenomenon and free **qua** noumenon. In adopting this strategy I do not wish to imply that the doctrine is not of philosophical importance or that Beck or I consider it unimportant. It is here left out of consideration for the simple reason that both of us reject it as unsatisfactory.[1]

The problem of effective freedom can now be put in the form of the following question: Is a voluntary agent ever correct in assuming that whereas the retrospect and his chosen bodily conduct predetermine the prospect, the retrospect by itself does not predetermine either the chosen bodily conduct or the prospect? The determinist's - or to be careful the intrawordly or intraphenomenal determinist's - view is that the assumption is never correct. This answer is, among many others, given by Leibniz who analyses causal determination in terms of sufficient reason. It is given by Locke and his empiricist successors, who derive it from the science of their day. And it is given by Kant - even though he also defends a concept of extraphenomenal or noumenal freedom which he regards as both indispensable (unentbehrlich) for the existence of morality and art and as inconceivable (unbegreiflich).[2] An indeterministic answer is given by some existentialists, who as a rule do not take much notice of science and also, which is here more important, by C.S.Peirce who argues that the thesis that science - including Newton's physics - implies determinism is based on a misunderstanding of its content and method.[3]

Both, determinists and indeterminists may, of course, agree on some concept of empirical freedom, defined as the absence of various kinds of coercion. Whereas, as Beck points out, the question as to whether or not an action was empirically free can be decided by empirical tests, no such tests are available to decide whether any action at all can be effectively free in the sense asserted e.g. by Peirce, and denied e.g. by Locke and his successors. For the determinist would have to show that to every impression that an action conforms to the conditons 1-3, there exists a falsifying correction, while the indeterminist would have to show that there exists at least one such impression which is not falsifiable. Although the issue between determinism and indeterminism is empirically undecidable it is not, as Beck clearly shows, immune to arguments.

2. Moral Freedom, Commonsense and Science

I am - as, I presume, are most people - under the impression of sometimes having the ability effectively to intervene in the course of nature (in a manner conforming to conditions 1-3). I am, moreover, convinced that this impression is sometimes correct. This conviction is in no way shaken by the great variety of deterministic arguments known to me, since all of them seem to me to be based on easily discoverable mistakes. In the present context it is sufficient to consider only two such arguments, namely the argument or rather the assertion that the assumption of moral freedom is unintelligible or, at least, inconsistent with commonsense and the argument that it is inconsistent with science.[4]

In order to show that a view of effective freedom is consistent with commonsense and intelligible to commonsense thinking, it is not necesary to hold that commonsense is unchangeable, that there is only one set of commonsense beliefs or that commonsense is the ultimate judge of philosophical theses. What is necessary, is to point to a sufficiently widely familiar region of non-philosophical thinking which is generally agreed to be intelligible - even if what is agreed to be intelligible is rejected on philosophical, scientific or other grounds. I find a concept of intelligible, effective freedom, which is, at least, very similar to mine and Beck's, in the Bible (a fact which is perfectly compatible with my being an agnostic) and also in Goethe's **Faust**.

Briefly, according to the begining of the book of **Genesis** God created man in his own image, which among other things implies that God gave man the power and the freedom to act morally or immorally and also the freedom to create, in particular, the freedom involved in artistic creation. Within God's power of creation, the Bible implicitly distinguishes two aspects, namely the power to create something out of nothing and the power to create order out of chaos. Whatever the power of a **creatio ex nihilo** may mean, it need not be ascribed to man **qua** moral agent and, one may add with Kant, **qua** creator of works of art. In order to explain these features of human existence, it suffices to ascribe to human beings some small measure of the biblical God's power of **creatio ex chaos** or **ex tohuvabohu**, i.e. of the power to impose some order on comparative chaos in a manner which is not determined. If man possesses this power, which is perfectly understandable to believing, atheist and agnostic readers of the Bible, then it makes sense to assume that not all his actions are predetermined, i.e. that the three conditons which always **appear** to be fulfilled by voluntary actions are sometimes actually fulfilled by them. In other words, it then makes sense to assume that man in some of his actions is not just free from coercion but "really", "genuinely" or "effectively" free.

The conception of man as free to act morally or immorally and as free to create by imposing his own order on comparative chaos, is central to Goethe's **Faust** and to the Western culture in the context of which it can be understood. It is man's freedom and creative power which makes him the little God of the world (**"der kleine Gott der Welt"**). That man can - independently of any explicitly philosophical point of view - be conceived in this way, shows the possibility that no illusion need be involved in an experience of freedom of choice which, as Beck puts it, "is **schlicht gegeben** and has a self-evidence greater than any causal law which would entail its necessary illusoriness." It is worth noting that Beck in the course of explaining his conception of freedom also refers to **Faust** by quoting and implicitly rejecting Mephistopheles' insincere **dictum** that man believes to push, whereas in fact he is always being pushed.

Among the arguments which are meant to show that my assumption of indeterministic freedom and, hence, the assumption of man's power to act as a little God, rather than a fully programmed machine, is unintelligible or false, the argument which is probably most often used is the alleged inconsistency of effective freedom with science. This argument is based on a failure to notice that scientific theories are not descriptions but idealizations, of the world of common experience or the **Lebenswelt**. They replace this world by a different world with which - for the purpose of the theories and in the context of their use - they can be identified. The difference between the logico-mathematical structure and the conceptual net of a theory on the one hand and of commonsense thinking on the other, implies that it is a gross error, to regard the theory as applicable beyond the limits of its identifiability with commonsense thinking. Thus Newton's theory, as Newton well knew,[5] describes a world consisting of nothing but particles possessing position and momentum and moving in accordance with the three Newtonian laws of motion. It is a world in which there is no room for human action - not even in a sense which would permit the distinction between action under coercion and action which is free from it. That for the purpose of applying Newton's theory and in the context of its application the possibility of human actions, in particular of creative human actions, is usefully ignored, does not imply their non-existence or - even worse - their impossibility. Similar remarks apply to other scientific theories, e.g. those which for their purpose replace the concept of man by a concept of **homo economicus, homo physiologicus** etc.[6]

3. A Brief Comparison of Beck's Account of Freedom With the Account Outlined Above

My fundamental agreement with Beck's conception of real or effective freedom is that, unlike Kant, he places it in the world of experience. The manner in which he shows this to be possible is connected with his distinction of several Kantian concepts of freedom, in particular with his illuminating exhibition of a very important change in Kant's later philosophy. Whereas, as Beck shows, Kant in the first Critique clearly regarded the principle of causality as a constitutive principle, he tends in the third Critique to change its logical status to that of a merely regulative principle. This change to some extent anticipates Beck's own postulational view of freedom, which bases the consistency of the principle of freedom and of natural necessity on their demotion from constitutive principles to complementary regulative principles, maxims or postulates. No inconsistency arises if, following Beck one accepts freedom as "a postulate for practical and moral life" and natural necessity as "a postulate for research into nature and human nature".

In making up my mind about the concept of effective freedom, I tried to weigh the following considerations against each other: (1) the conviction of being free, based on introspective evidence and the tendency to accept and apportion moral blame; (2) the content of contemporary science, which is no longer deterministic - even though an indeterministic science does not imply either the existence or the non-existence of free will; (3) the general character of science, especially mathematized science, as idealization rather than as description of the world of experience (see above); (4) the availability of reasonable systems of logic, which do not contain the law of excluded middle and thus do not imply that a proposition stating the occurrence of any event **at a certain time** is - non-temporally or eternally - true or false; and last (5) the intelligibility of a commonsense concept of freedom which in its original form or after some philosophical refinement is, in Kant's words, the **causa cognoscendi** of morality and also, it may be argued, of human creativity.

The first and the last consideration support my assumption of effective freedom, while the others - at the very least - do not weaken it. I can, therefore, briefly express the difference between Beck's and my own view in Kantian terminology by saying that whereas Beck regards natural necessity and freedom as postulates, I regard the former as a (possibly obsolete) postulate and the latter as a constitutive principle of being a person. Yet this difference between our accounts is, as I said at the beginning, far outweighed by their agreement on what I take to be their most important consequences, namely the location of real or effective freedom within the world of experience and the rejection of psychological determinism which Beck, following Sartre, describes as "an attitude of excuse".

* * * * * * * * * *

FOOTNOTES

1. For a more detailed discussion of the concept of a voluntary action, see **Experience and Conduct** (Cambridge University Press, 1976) Chapters 5 & 6.

2. See **Critique of Practical Reason,** Preface and **Critique of Judgment,** Section 43.

3. See "The Doctrine of Necessity Examined", **Monist** Vol.2 (1892) and **Collected Works** (ed.) Hartshorne & Weiss (Harvard University Press, 1960) 6, 36-65.

4. The latter argument has been considered e.g. in **Fundamental Questions in Philosophy** (London, 1971) Ch.XVI and will therefore be discussed with the utmost brevity.

5. For Newton's views see e.g. Richard S.Westfall "Isaac Newton's 'Theologiae Gentilis Origines Philosophiae'", in **The Secular Mind,** (ed.) W.W.Wagar (New York, 1981).

6. For a more detailed discussion of the relation between commonsense and theoretical thinking, see e.g. **Experience and Theory** (London, 1966).

* * * * * * * * * *

Srzednick, J. T. J., Stephan Körner – Philosophical Analysis and Reconstruction. ISBN 90-247-3543-2.
© 1987. Martinus Nijhoff Publishers, Dordrecht. Printed in The Netherlands.

4. THE MODES OF PHILOSOPHICAL INVOLVEMENT WITH A CATEGORIAL FRAMEWORK

WITOLD MARCISZEWSKI

Stephan Körner in an extensive essay, has convincingly shown the importance of what he calls "categorial frameworks" in our understanding of the world.[1] The aim of the present essay is to contribute to a better understanding of the very idea of a categorial framework.

The choice of a categorial framework, that is a system of ontological categories, is the first step in a philosopher's self-determination. Körner's essay helps those who are looking for such a determination. They are provided with a list of questions each of which is to be answered by "yes" or "no"; the list, contained in Chapter I (p.4) is what the author calls "the phases of categorization".

We ask here whether this list of questions alone - before any answers are given - is philosophically indifferent or it whether it requires a mode of philosophical involvement. It may be expected that **metaphilosophical** investigations, like those of Körner, do not involve a specific philosophical position. On the other hand, Körner's list of categorial questions seems to be philosophically involved (though much less involved than any list of answers). Hence the point requires an examination.

From among seven phases of categorization listed by the author, all but the first are stated as disjunctions of the form "p or not-p".[2] Such a disjunction can be restated as a whether-question "p?". For instance: whether the class of dependent particulars is empty or not? This sentence exemplifies the typical form of question occurring in the category register. When seeing it as a disjunction, we can put it in the following symbolic form:

(a) $\{x: F(x)\} = \phi \lor \{x: F(x)\} \neq \phi$.

1. On the Existence of Categories as Classes

Formula (a) does not assert the existence of members for the class $\{x:$

$F(x)$} but it implies the existence of the class itself by virtue of the axiom of comprehension:

(b) $\exists y \forall x (x \in y \leftrightarrow F(x))$,

the y in question being the class {x: $F(x)$}.

The use of the comprehension axiom should be secured against antinomies, e.g. by the following modification of (b), called the axiom of subsets:

(b*) $\forall z \exists y \forall x (x \in y \leftrightarrow (x \in z \ \& \ F(x)))$;

then the existence of a class z including y has to be assumed as well. Hence at least two assumptions concerning the existence of specific classes are associated with the disjunction (a). To what extent are such assumptions philosophically committing? Since in our discussion the class {x: $F(x)$} forms a category, e.g. the category of dependent particulars, the existence of the respective category is ascertained in virtue of (b) or (b*). What may be rejected is a partition into categories, namely a partition in which a member is empty, since a correct partition is a family of non-empty classes.

Thus we have the first reply to the question whether Körner's register of categories commits us to a philosophical decision. The decision involved is concerned with the existence of classes, especially the empty class which by no means can be construed as a mereological whole.

Let us suppose, however, that our speaking of empty classes is merely a **modus loquendi**, and, instead, we can speak of the non-existence of the objects in question, i.e.:

(c) $\neg \exists x F(x)$.

When this modification of the manner of speaking is accompanied by the refusal to accept the axiom of comprehension, then any involvement in the existence-of-classes theory is avoided.

There is, however, another involvement which holds both whether we speak of the emptiness of the class {x: $F(x)$} or of the non-existence of F's. In either case the meaningfulness of the predicate F is taken for granted, at least tacitly. Then, even if a category does not exist as an extension (a class), it does "exist", so to speak, as an intension; and holding that is also a mode of philosophical involvement (see Sect.3 of this paper). This mode of commitment can be demonstrated as follows. Either the predicate F is primitive or defined. If it is primitive, it must have been introduced by a set of meaning postulates, and those, being assumed to be true philosophical propositions, give rise to philosophical involvement. If F is a defined concept, then there are defining concepts which again require meaning postulates.

The only way to avoid such an involvement is to prevent entering F into our vocabulary. It is an admissible strategy, but a costly one, as it ruins the dialogue between those who accept the existence of classes such as {x: F(x)} and those who refuse to accept the existence of any classes. There is hope of success for such a dialogue, but only on the condition that the parts of the dialogue have a common language.

2. On the Existence of Attributes

The first item in Körner's category register is the partition of all objects into a class of particulars and a class of attributes; unlike the remaining items, this one is expressed in a categorical, and not in a disjunctive statement (the latter admitting either the acceptance or the rejection of a partition). Hence it constitutes an assertion, not a question. This implies a further philosophical involvement, viz. the assertion of the existence of attributes. Two questions follow from this assertion. First, is it possible to avoid this additional involvement without damaging the construction created by Körner? Second, are there any reasons to try that?

It is easy to see that the answer to the first problem is in the affirmative. It is enough to turn (i) into the following disjunction (cp. note 2). "**Either** acknowledgment of a partition of all objects into (a) a class of particulars, i.e. objects which are logically ultimate and (b) a class of attributes, i.e. objects which are not logically ultimate; **or** rejection of such a partition because one of the classes is empty."

As regards the second question, whether it is worthwhile to restate (i) in this way, the answer depends on how seriously we take the nominalistic arguments against attributes. Körner seems not to notice these arguments as he writes as follows: "As regards (i), the partition of all objects into particulars and attributes is generally acknowledged. The clearest modern version of it is probably that of Frege" (p.4). I do not think that this passage expresses an historical truth. The partition in question happened to be rejected, even vehemently rejected, by such nominalists as Tadeusz

Kotarbiński and some Polish authors following him. It looks so not only from the Polish perspective. Paul Weingartner (when discussing various universes proposed for metaphysics) considered what he called a Parmenidean universe, i.e. that admitting of just one category of objects, namely the category of particulars.[3]

Thus the partition of all objects into particulars and attributes does not prove generally acknowledged. However, it can be claimed that this partition **should** be generally acknowledged. A reason for this demand is hinted at in (i) by the phrase "logically ultimate" as referring to attributes. In the items following (i), viz. (ii) and (iii), the opposition between the members of a partition is stated not in terms of a logical quality but in the terms "ontologically fundamental" and "not ontologically fundamental".

The term "logically" in the context of (i) is relevant for the following reason. The basic relation in logic is that of predication; it holds even at the lowest level of syntactic constructions, viz. the level of atomic sentences. The syntactic relation between the predicate and the grammatical subject of a sentence has its objective counterpart in the relation between an attribute and a particular. Furthermore, in the symbolic language of logic this distinction is rendered by distinguishing two syntactic categories of expressions: individual variables and predicate variables. Individual variables are used as grammatical subjects and are logically ultimate, in the sense that they cannot be predicated of anything. Therefore their extralinguistic counterparts, viz. particulars, can be also called logically ultimate.

Now it can be seen why the partition into particulars and attributes should be generally acknowledged, provided that there is correspondence between our language and the external world. The correspondence has to appear at the most fundamental level; that is, the structure of the simplest sentence must be a mapping of an elementary structure into reality; and this mapping involves both particulars and attributes.

However, let us suppose that our nominalistic opponent perseveres in holding that only the logically ultimate expressions can have their counterparts in reality, hence attributes do not exist. Can we refute such a claim? If we let our opponent express his point in the language of logic, then he might obtain the following statement:

(NA) $\neg \exists A$ (A is an attribute).

This, however, is not feasible if he happens to be guided by Quine's principle "To be is to be the value of a bound variable", because the very fact of binding a predicate variable (like A in the above formula) involves him in the existence of attributes. Now the nominalist has two options: either to give up Quine's maxim and to accept (NA) with all its consequences, or to look for a way out in metalinguistic considerations. The discussion of the later option will be postponed to Section 3.

Then, assuming the first option, we proceed by asking our nominalistic opponent whether he accepts the Leibnizian conception of identity of particulars as expressed in the formula:

(L) $x = y \leftrightarrow \forall p \in A$ (x has p $\leftrightarrow$ y has p),

where A is the set of all attributes. If he does not accept (L), he will be obliged to propose his own definition of the identity of particulars; if, looking for such definition, he resorts to the metalanguage strategy (e.g., x and y are identical if whatever is truly said of one of them is true of the other too) - again we postpone the discussion to Section 3. If he agrees with (L), we argue as follows. The restriction under the quantifier yields the antecedent of an implication whose consequent is: x has p $\leftrightarrow$ y has p. If the nominalist is right, then the right side of (L) is always true because of the false antecedent. Then, it is true idependently of what x and y are; they may be, for instance, Caesar and Brutus. Then, the right side of (L) being true of any pair of particulars, the whole (allegedly) true (L) renders them identical, as then the left side must be true. Hence, there is at most one particular in the world.

In other words: if one denies the existence of attributes, it amounts to saying that in every sentence nothing is predicated about something; hence the same (i.e. nothing) is predicated of all particulars; hence all particulars are identical; hence there is only one particular.

In the face of so untenable a conclusion the nominalist may turn to the other option. It will consist in placing attributes in a "world" very different from that reality in which particulars exist.

3. On the Giveness of Noemata

There is an important distinction, proposed by Roman Ingarden, which proves especially useful for the present discussion.[4] It is the distinction between two kinds of philosophical activity; Ingarden called them "ontology" and "metaphysics" giving each of these terms a different meaning (though they are used interchangeably by some authors). Ontology, roughly speaking, is a study of the contents of our thoughts with the aim of distinguishing those which are internally consistent; it can be called the study of pure possibilities, or possible worlds. The classes of possible worlds being the subject matter of ontology may differ with each other as regards their categorial frameworks.

Metaphysics in the Ingardenian sense should settle the question which of the possible categorial structures is the structure of the real world; or, when speaking in Körner's terms, which of the categories listed in his register is non-empty (provided this list is complete). For example, if one admits the category of independent attributes (e.g. Plato's forms) as not being internally inconsistent, then he provides an ontological solution. If, in turn, he recognizes this category as non-empty, then he settles a metaphysical question.

This Ingardenian distinction parallels in a way the Husserlian distinction between **noema**, i.e. the content of intentional experience (according to Webster's definition), and the object referred to (by means of this content); let the latter be called the **referent**, after a paper by G.Küng.[5] The notion of noema is related to the distinction between act and content of thought as stated by F.Brentano and K.Twardowski. The acts themselves belong to the domain of psychology. Their contents, i.e. noemata, if concerned with philosophical questions, belong to the domain of ontology. The referents of such noemata, if there are any, belong to the domain of metaphysics. Hence, to decide whether certain noemata have referents or not is a metaphysical enterprise (as for other noemata, it is the task of empirical sciences). For example, it is a metaphysical question whether the noema expressed by the phrase "independent attribute" has a referent or not; if not, then the phrase denotes the empty class. It can be said about referents that they actually exist, and about noemata that they intentionally (mentally) exist, or, better, that they are given to minds.

Moreover, as Küng observes, there is a parallel between the concepts of noema and referent and the Fregean concepts of sense (**Sinn**) and reference (**Bedeutung**), respectively. In modern logic, an alternative to Fregean semantics originated with Russell, the notion of sense does not appear. Let me use a quotation from Küng's paper to give an account of this approach. "A realist like Russell is more interested to correlate the expressions with counterparts in reality than to assign them beings of reason as their senses. He will, for instance, stress that predicate expressions must refer to (or designate) properties or relations in reality rather than worry about the conceptual meanings which are their senses".

Thus, Russell and his followers tell us there is little use in dealing with senses in semantics, but they do not tell us that there is no sense at all in speaking about senses. There is a more radical approach, that of the behaviorists who found a philosophical advocate in G.Ryle (in **The Concept of Mind**). They claim there are no such things as mental acts; hence there are no contents of mental acts. Hence no senses.

The appearance of referential semantics Russell's style, and the appearance of the theory of non-existence of minds bear upon the theory of noemata and its degree of philosophical involvement of the theory of noemata. For if diverging views are possible, the choice between them is more involving than if there were only one possibility. True, we can disregard an option as not-conforming to our standards of intelligibility (e.g., for the present author the behavioristic view is unintelligible), but then we need, as Körner points out, a categorial framework providing us with such standards. Hence, if we are to avoid a **petitio principii**, in discussing metaphilosophically the philosophical involvement of a categorial framework (i.e., a philosophical theory), we have to content ourselves with the historical (or social) notion of involvement mentioned above.

From this point of view Körner's list of categorial questions is not free of a philosophical involvement. Even if we dispense with attributes (by the modification considered in Sect.2) and with classes (due to a nominalistic strategy), there remains the question of whether a category is empty or not. This means that we have the idea of category before being acquainted with its members, hence the idea must have entered our language **via** to a noematic experience; when acknowledging this fact we endorse the noemata theory.

Noematic semantics offers a convenient strategy for nominalists. For instance, they can deny the existence of attributes even by means of the formula

$\sim \exists A$ (A is an attribute)

provided that the variable A ranges over a domain of noemata. Possibly this solution resembles the substitution interpretation of quantifiers which, in turn, is similar to Leśniewski's conception of quantifiers.[6] If a nominalist is ready to accept the giveness of noemata, at this price he can avoid any other philosophical involvement. He can, for instance, construe attributes and classes as mappings of noemata into the domain of (actual) particulars, the empty class corresponding to the noema expressed by "$x \neq x$". Such a Berkeleyan version of nominalism can be alien to some modern nominalists, but there seems to be no other way out. Everyone then must decide whether he prefers to get involved with the domain of noemata or to get dogmatically involved with the existence of attributes, relations, classes etc. before he can adduce any metaphysical arguments supporting his decision (for metaphysics presupposes ontological inquires into the domain of noemata).

Körner's choice seems to be in favour of a noematic world. Thus, he can safely speak of categories which may prove either empty or non-empty, according to possible later considerations. Then, should he express his propositions in symbolic formulas, he would not be bound by Quine's principle. Thus, although his metaphilosophcial theory of categorial frameworks is philosophically committed, the involvement is less than that found in any alternative approach.[7]

* * * * * * * * * *

MODES OF PHILOSOPHICAL INVOLVEMENT

FOOTNOTES

1. Stephan Körner, **Categorial Frameworks** (Oxford, Blackwell, 1970).

2. Here is the passage in question - from page 4.
   The phases are (i) Acknowledgement of a partition of all objects into (a) a class of particulars, i.e. objects which are logically ultimate and (b) a class of attributes, i.e. objects which are not logically ultimate. (ii) Either acknowledgment of a partition of all particulars into (a) a class of independent particulars, i.e. particulars which are ontologically fundamental and (b) a class of dependent particulars, i.e. particulars which are not ontologically fundamental; or rejection of such a partition because one of the classes is empty. (iii) Either acknowledgment of a partition of all attributes into (a) a class of independent attributes or universals, i.e. attributes which are ontologically fundamental and (b) a class of dependent attributes, i.e. attributes which are not ontologically fundamental; or rejection of such a partition because one of the classes is empty. (iv) If the class of independent particulars is not empty, then either acknowledgment of a partition of this class into two or more maximal classes of independent particulars or rejection of such a partition as unnatural (e.g. because there exists only one independent particular). (v) If the class of dependent particulars is not empty, then either acknowledgment of a partition of this class into two or more maximal classes of dependent particulars or rejection of such a partition as unnatural. Lastly, (vi) and (vii) which read respectively like (iv) and (v) with 'attribute' taking the place of 'particular'.

3. Paul Weingertner, "Das Poblem des Gegenstandbereiches in der Metaphysik", **Salzburger Jahrbuch für Philosophie** XIX (1974), pp.35-70.

4. Roman Ingarden, **Der Streit um die Existenz der Welt** vol.1 **Existentialontologie** (Tübingen, Niemeyer, 1964). See pp.33 and 188.

5. Guido Küng, "The World As Noema and as Referent", **Journal of the British Society for Phenomenology** Vol.3 (no.1, January 1972) pp.15-26. The paper brings very useful historical information and a thorough explanation of the notions mentioned in its title.

6. See G.Küng and J.T.Canty, "Substitutional Quantification and Leśniewskian Quantification", **Theoria** 36 (1970) pp.165-182.

7. Already after submitting this paper to the Editor, I found a thorough examination of the same Körner ideas contained in the study "Logic and Ontology in the Study of Theory Change" by Stig Andur Pedersen (Copenhagen); the study appeared in **Poznan Studies in the Philosophy of the Sciences and the Humanities** Vol.3 (nos.1-4 1977) (B.R.Grüner Co., Amsterdam). Pederson, basically agreeing with Körner's approach, suggests a restatement of condition (i) - quoted above in footnote 2 - to transform it into the disjunction as suggested by me at the start of

Section 2 in the present paper (see p.49 in the quoted issue of **Poźnan Studies**). For his claim Pederson offers a convincing argument, referring to the set-theoretical axiom of regularity; thus his argumentation is different from that presented in my paper. Pedersen discusses still other views of Körner's on the role and structure of categorial frameworks, and hence his contribution can be recommended to all those who are seriously interested in the core of Körner's ideas.

* * * * * * * * *

REPLY TO PROFESSOR MARCISZEWSKI

Stephan Körner

Professor Marciszewski's discussion of my conception of the structure and function of categorial frameworks, as propounded in my 1970 monograph, shows a full and, I am pleased to note, sympathetic understanding of the task which I set myself. It also contains some constructive and justified criticisms. Indeed some of the modifications proposed by him will be found in a book of mine which I submitted to the Cambridge University Press, before I had the benefit of studying his comments.[1] My aim in both these books, as Marciszewski clearly sees, is not to develop and to defend my own metaphysics or, more particularly, my own categorial framework i.e. the supreme principles governing my thinking about what I take to be the world of intersubjective experience. It is to develop the general notion of categorial frameworks, of which my own is one example among many. Any attempt at fulfilling this task is exposed to the everpresent danger of confusing features peculiar to one's own categorial framework with features characteristic of any such structure. While I do not claim to have been successful in avoiding this danger, I do claim that I have been fully aware of it. This awareness finds its probably clearest expression in the last chapter of my later book, which contains a brief synopsis of the convictions which constitute my immanent and my transcendent metaphysics as well as my morality. The purpose of the chapter is to enable the reader to judge for himself, how far I have been able to protect my metaphilosophical inquiry from involvement with my philosophical convictions.

My reply to Marciszewski's comment falls into two parts. In the first I shall briefly explain the nature of my inquiry and the status of the theses in which it should result, if correctly pursued. What I have to say in this part has either found Marciszewski's explicit approval or is, I think, implicitly approved by him. In the second part I shall deal with his charges of illegitimate philosophical involvement.

STEPHAN KORNER

1. On the Concept of a Categorial Framework as a Tool of Metaphilosophy

The notion of a categorial framework is meant to make the notion of immanent, as opposed to transcendent, metaphysics more precise. The principles defining a person's categorial framework are supreme in the sense that any proposition which is incompatible with such a principle must be rejected by him. The principles defining a categorial framework thus remind us of Kant's synthetic principles **a priori**, in particular his anticipations of perception and analogies of experience, as well as of Collingwood's absolute presuppositions. They differ from the former, mainly in not implying the claim of being unique i.e. in excluding the possibility of alternative categorial frameworks. And they differ from the latter mainly in being regarded as propositions which are true or false.

While in **Categorial Frameworks** I characterized a categorial framework as the result of a categorization consisting of a list of questions asked in a certain order and requiring an affirmative or negative answer (apart from possible border-line cases), I now regard it as the result of the manner in which human beings as a matter of empirical fact organize their experience of the inter-subjective world. The organization which is explained in the later book is, I think, more adequate and less restrictive and may thus result in categorial frameworks which are more likely to find Marciszewski's approval not only in principle, but even as regards their detailed characterization. Yet, however this may be, the comments he makes are still highly instructive and worth answering.

Before trying to do so, it will be helpful to say a few words about the relation between on the one hand the concept of a categorial framework as a tool used in the inquiry into the nature of metaphysical thinking and on the other hand metaphysical thinking itself. This can for our present purpose be done by means of a fairly simple example. If we imagine a contemporary of Kant to have been engaged in trying to characterize the systems of immanent metaphysics accessible to him through conversations with his contemporaries, historical documents and his own philosophical thought experiments, one would, it seems to me, not be surprised if he characterized **all systems** of immanent metaphysics as containing among their supreme principles the principles of Aristotelian logic (allowing for minor modifications) and a principle of continuity to the effect that **natura non facit saltus** or - more precisely - that being a changing entity which, in spite of changing, preserves its identity, logically implies that the change is continuous.

It is important to distinguish between the principles of classical logic and the principle of continuity on the one hand and our metaphilosopher's thesis that these statements constitute a common (constant, unchangeable, indispensable) part of any immanent metaphysics known to or conceivable by him. As it stands, the latter thesis is an empirical or, more precisely, an anthropological statement about the manner in which human beings - as opposed to subhuman or superhuman beings - think about the world. The statement is empirical in the sense that, like other anthropological statements, it is in principle refutable - whether it has been in fact refuted or not. As against this, the principles of classical logic and the principle of continuity (e.g. in its Leibnizian or Kantian version), the acceptance of which by all or some human beings, forms the subject of the empirical metastatement are non-empirical. Our metaphilosopher is thus making an empirical statement about one or more non-empirical statements. This is so even if he yields to the temptation to regard his metastatement as non-empirical and in some sense as necessarily true. Among the philosophers who have yielded to this temptation are Descartes, Husserl and Brentano, who regard the metastatement as being self-evidently true, as well as Kant and some neo-Kantians who regard it either as a transcendental, synthetic proposition **a priori** which can be proven by a transcendental deduction or else as a somewhat disguised analytic proposition.

While I find it understandable that our metaphilosopher regarded his metastatement as unlikely to be refuted, his claim that it is irrefutable and any alleged justification of the claim must be mistaken - for the simple reason that some later systems of immanent metaphysics no longer contain the principles of classical logic or the principle of continuity. Examples are von Neumann and those of his followers whose reflection on quantum mechanics have led them to the acceptance of a supreme principle of discontinuity and of a logic which radically deviates from the classical.

My concept of a categorial framework has been formulated as a result of inquiring into the manner in which human beings **as a matter of empirical fact** organize their thinking. It is intended to serve as a tool for further inquiry, which may well reveal that it is not adequate to its task, e.g. because the generic notion of a categorial framework is too wide, too

narrow or otherwise defective. The reasons for this inadequacy may in
particular be a neglect of ways of thinking which are removed from one's
own or putting too much weight on one's own way of thinking. Ideally there
should be an interaction between the inquiry and the tool by means of which
it is being pursued. While the results should be improved by the use of the
tool, the tool should be improved in the course of the inquiry. Both kinds
of improvement are furthered by recognizing that in stating that human
thinking proceeds under the constraints of a categorial framework, one is
neither stating the logical consequences of a definition of 'thinking' nor
a transcendentally necessary condition of thinking. Both types of statement
protect the notion of a categorial statement from criticism. If however the
notion of a categorial framework is regarded as an improvable tool, used in
an empirical inquiry, then such protection is undesirable and criticism
welcome.

2. On Marciszewski Criticisms

Marciszewski rightly points out that in **Categorial Fameworks** I seemed
to take the notion of classes, in particular the notion of the empty class
for granted. He rightly assumes that I did not wish to burden my account of
categorial frameworks with an antinominalist thesis and that my intentions
are best understood if what I say about classes and might have said about
the empty class is no more than a **modus loquendi.** He further rightly points
out that in this case a dialogue between "those who accept the existence of
classes... and those who refuse to accept the existence of classes" must be
admitted as possible and made possible by providing the opponents with a
common language.

In my later book I distinguish between particulars and attributes - the
particulars being instances of attributes and the attributes determining a
range or aggregate of particulars. I draw attention to the differences between the class-element relation which is made precise in various versions
of class-logic and the whole-part relation which is made precise in various
versions of mereology. But I argue that the differences may not only be
overlooked, but may for some human beings be irrelevant or non-existent.
Such people may be using a notion of aggregate, which does not imply a
distinction between class and whole, a notion of component which does not
imply a distinction between element and part and a component-aggregate

relation which does not imply a distinction between the element-class and the part-whole relation.

Having acknowledged the possibility of a neutral component-aggregate logic I turn my attention - perhaps prematurely - to philosophical issues which require a decision as to whether an entity is a class of entities or a whole, e.g. whether space is a class of relations, as held by Leibniz, or a whole consisting of spatial parts, as held by Kant. Had I had the benefit of Marciszewski's remarks before submitting my later book, I should have discussed the possibility of a class-less nominalism at greater length. Indeed while my own metaphysical convictions are antinominalist for reasons which are very similar to those which Marciszewski adduces in the later part of his second section, I would in pursuing my metaphilosophical inquiry try to do my best to defend the possibility of a nominalist, immanent metaphysics.

Marciszewski thinks that in **Categorial Frameworks** I disregard a distinction proposed by Roman Ingarden, namely the distinction between the study of pure possibilities or possible worlds, which Ingarden rather idiosyncratically calls "ontology" and the study of the structure of the real world which he no less idiosyncratically calls "metaphysics". My reply to this is that I not only acknowledge the distinction as important, but that I am trying to improve on it by rejecting the assumption made by Ingarden and, before him, by Kant and Leibniz that there is one and only one **logically** possible conception of a logically possible world, because there is one and only one logic in terms of which logical consistency and, hence, logical deducibility as well as 'being a possible world' are defined.

As I make clear in **Categorial Frameworks** and, I think, even clearer in the later book, every categorial framework is embedded in a logic which need not be the same for all categorial frameworks. (For a discussion of various logical systems and their relations to each other see the Appendix of **Categorial Frameworks** and for a more detailed account Chapter 5 of the later book.) It will be useful to recall that Leibniz distinguishes between a merely possible world which is subject to the principle of contradiction and the real world which in addition is subject to the principle of sufficient reason; and that Kant distinguishes between formal logic (more precisely, a slightly modified version of Aristotle's formal logic) which determines what is logically possible and transcendental logic which determines what is materially possible. In a similar way I distinguish between on the one hand the logic which underlies a categorial framework and which determines what is logically possible, i.e. expressible by propositions which are by virtue of this logic internally consistent; on the other hand the supreme principles which determine the (maximal kinds of the) categorial framework and what is materially possible i.e. expressible by propositions which are consistent with these principles. My own view thus differs, **roughly speaking**, in two important respects from Kant's. First, whereas Kant admits only one set of categories, the applicability of which determines what is possible within his categorial framework, I admit a plurality of mutually inconsistent sets of categories and of categorial frameworks. Second, whereas Kant admits only one logic, I do admit a plurality of logical systems, e.g. the standard classical and the standard intuitionist logic. For me the locution "the set of all possible worlds" is elliptical and has to be replaced by "the set of all possible worlds as demarcated by the logic L", where L may be a classical or constructivist, an exact or inexact logic etc. (See my later book).

Since Marciszewski in his comments expresses some misgivings about my admitting the possibility of empty categories it seems desirable to indicate, how the notion of emptiness of a concept could be made more precise. At the moment it would seem to me best to distinguish between three kinds of emptiness, namely logical emptiness, categorial emptiness and factual emptiness. For the acceptor of a categorial framework determined by an underlying logic L and supreme non-logical principles F a concept P is logically empty if, and only if, its applicability is incompatible with L, categorially empty if, and only if, its applicability is in accordance with L incompatible with F; and factually empty if, and only if, it is not applicable, although it is neither logically nor categorially empty. (My later book contains similar distinctions between various kinds of meaninglessness in Chapter 13, Sect.4). In elaborating these distinctions, I should point out that Marciszewski's concept of a logically possible world is more restrictive than mine, since his preference for a version of classical logic finds expression not only in his philosophy, but also in his metaphilosophy.

FOOTNOTE
1. See **Metaphysics: Its Structure and Function**, (Cambridge, 1984).

* * * * * * * * * *

Srzednick, J. T. J., Stephan Körner — Philosophical Analysis and Reconstruction. ISBN 90-247-3543-2.
© *1987. Martinus Nijhoff Publishers, Dordrecht. Printed in The Netherlands.*

5. ESTABLISHING THE CORRESPONDENCE THEORY OF TRUTH AND RENDERING IT COHERENT

Richard Sylvan

In his rewarding book, **Fundamental Questions of Philosophy**, Stephen Körner introduces the correspondence theory of truth as seemingly 'the most natural analysis of truth', explains what the theory tries to do, and what it does **not** attempt, defends the theory against some standard objections, and advances difficulties for rival theories of truth (see p.101ff.). What follows removes the limitations Körner claims to find on the correspondence theory; it also elaborates upon what Körner has explained, but at the same time makes one or two significant variations, as will appear.

1. Correspondence and Facts

According to the correspondence theory truth corresponds to the facts, or, in a first expansion, what is true corresponds to what is a fact. That is, there is a correspondence (indeed, more than this, an isomorphism) between true statements and facts. Many are the philosophical pundits who have claimed that the correspondence theory is mistaken, that there is no such correspondence, that it can't be done.

The pundits are wrong. The argument that they are mistaken is so simple that it is a bit surprising it hasn't been presented dozens of times before. But the obvious has a way of eluding philosophers. The argument to the correspondence theory is in essence simply this: just as a general Tarski theory of truth can be provided for every language (through the techniques of a universal semantics), so a general Tarski theory of facts can be provided in precisely the same way. But these theories are isomorphic, which provides the sought correspondence.

A typical base clause of the analogous truth-facts theory will illustrate how the correspondences go:

⌜aRb⌝ is true iff aRb
          iff ∫aRb is a fact

where the squiggle ∫ reads, as often, 'that'. At the base level, truth corresponds to the facts, so it is usually conceded. But the rest of the story merely consists in proceeding to a recursive definition of factuality, in a way isomorphic to the Tarski-style definitions of truth. There are two ways of doing this, deploying positive and negative facts (as truth and falsity) or using negation (as truth and not) - ways that amount to the same in a two-valued setting, given an appropriate base to start from, **except** ontologically. Let us take the second way. The recursion clauses for quantification theory proceed naturally as follows, for connectives ~("not") and & ("and") and quantifier U ("every"):

    ∫~A is a fact iff ∫A is not a fact ["negative facts"]

    ∫A&B is a fact iff ∫A is a fact and ∫B is a fact ["conjunctive facts"]

    ∫(Ux)A(x) is a fact iff ∫A(y/x) is a fact for every y ["universal facts"]

There are of course variant forms that the clause for universal facts can be given, depending upon preferred account of quantification (standard substitution notation is presupposed).

To go beyond quantification theory the modelling apparatus is enlarged, just as for truth definitions. For instance, intensional functors are included in the general theory of factuality by incorporating worlds, or some equivalent, appropriately into the modelling apparatus. And so on, for the rest of full free λ-categorial languages with context (see e.g. US.). An evident theorem, **the correspondence theorem**, is this:

    for any A, ⌜A⌝ is true iff ∫A is a fact;[1]

what is true (statementally) are all and only what are facts.

The correspondence theory has been presented here in one of its factual outfits. But it could have been presented in a different factual garb. For example, as a first variation, replace 'is a fact' by 'belongs to the facts' or even (for a double correspondence form) 'corresponds to the facts'. And it could equally well have been given in its original Aristotelian form, or in Meinong's objective garb. For Aristotle, replace 'is a fact' by 'is the case', and for a full version replace ⌜∫A⌝ by ⌜†A⌝, which reads, ⌜what A expresses as being the case⌝. Then by the correspondence theorem,

⌜A⌝ is true iff †A is the case,

i.e. iff what A expresses as being the case is the case.
For intensional statements we have already presupposed pretty much the supplementation of Aristotle's theory suggested by Leibniz (see Körner's discussion p.101). For Meinong, replace 'is a fact' by 'is an objective', and replace ⌜∫A⌝ by ⌜ΠA⌝, where 'Π' is the objective transformation taking sentence into gerundive (being) form, e.g. ⌜a is β⌝ into ⌜a's being β⌝. Whereas Aristotle's theory is a static substance one, based on **cases**, Meinong's is closer to Whitehead's dynamic process position, which can also be approximated in this general framework.

The correspondence theory advanced differs from that Körner defends in only one respect, namely it claims to work for all truth, not merely for **'non-logical** truth' - a restriction Körner introduces (p.102) but does not use. No doubt he felt some unease with logical facts or cases, in contrast, for example, to universal ones.

This matter does touch however on a standard objection to the correspondence theory that Körner does not address, namely the objection from the excessive ontology the theory appears to some to involve. Even those prepared to concede the existence of simple facts, as in such prototypes as that **the cat sits on the mat**, are liable to become progressively more worried as the elaborate hierarchy of facts builds up through disjunctive facts, complex facts, universal facts, logical facts, intensional facts, and so on. The theory of objects, in terms of which the correspondence theory presented has implicitly been set, can break this ontological objection, by simply denying that facts exist, and also that they cannot therefore play a satisfactory role in a theory. Facts take on the role of useful theoretical objects both in philosophical theory and as adjuncts to commonsense. The claim (argued in JB) is that facts, like their mates propositions, do not have the right sorts of properties to exist. Applied to facts the existence terminology is, in any case, strained and otiose; people, apart from those touched by philosophy, don't normally talk of the existence of facts or need to do so. But because they aren't commonly said to exist, and don't exist, it doesn't follow that there aren't predicates which properly apply, or that they can't be integral to "the" world (or rather to suitably rich worlds). Facts are the case, positive facts obtain, and so on. More important, facts are part of the world; the world at any time is a knitting together of facts such as that **the cat sits on the mat**. This emerges from a larger account of the notion of **world**.[2]

Philosophers who want to see a nicely extensional world are going to take exception to the inclusion of such facts as that **2+2 is necessarily 4**, and worse, in the world. Well, we can of course distinguish purely extensional worlds, and perhaps sometimes there is point in doing so; but for larger enterprises these prove inadequate, and restriction to them results in much that is bizarre, as for instance Davidson's one Great Fact, reflecting a loss of appropriate discrimination of facts.[3]

2. Coherence Classically

Körner is surely correct in his criticism of such rivals to the correspondence theory as the self-evidence theory of truth (pp.104-5),[4] and also in much of what he argues against coherence theories, which he formulates in an exceptionally clear way (pp.105-7). But there is a remarkable gap in what he has to say regarding coherence theories - a gap which readmits these theories, though in a fashion that will make few philosophers happy, other than the synthesizers.[5]

Körner explains coherence in terms of a so far unanalysed relation of **involvement** on propositions of the same type as implication, a relation standing in for 'an obscure non-logical yet necessary connection between propositions', the notorious relation of **internal relation**.[6]

> In terms of the notion of involvement we can define the concept of a 'coherent set of propositions' and of a 'true proposition' and state the fundamental thesis of the coherence-theory of truth. A set of propositions is coherent if, and only if, (a) every proposition of the set involves every other proposition of the set, (b) every proposition which is involved by a proposition belonging to the set belongs itself to the set. A proposition is true - in the sense of the coherence theory - if, and only, it belongs to a coherent set. The fundamental thesis of the coherence theory is: there is one and only one coherent set of propositions (p.105)

Körner proceeds immediately to sketch the Hegelian archetype of the theory, and then to explain whence the theory derives its plausibility.

The requirements on coherence given are, in effect, strengthenings of the requirements, often given for coherence, of consistency and comprehensiveness, requirements which have appeared integrated in logical theory through the notion of maximal consistent sets with closure under very strong (two way) involvement requirements, and such, if the fundamental thesis is to be satisfied, that uniqueness results. It will be obvious already to logicians how this can be pulled off - in a remarkably non-Hegelian unidealist fashion.

'The difficulties to which the coherence theory of truth in every version (known to me) give rise', Körner explains, 'all stem from the notion of 'involvement''. The relation is undoubtedly some sort of implication, a relation theorists 'could easily explain if they were not prevented from doing so by the pressure of more important work' Körner adds. 'But' he continues, in what is supposed to be the main indictment of coherence theories, 'there is no known notion of ... implication ... such that any true proposition implies any other' (p.106). Isn't this a major gaffe? Körner tries to retrieve his position in the next paragraph (though seemingly as something of an afterthought):

> Ironically, there is one "if-then" relation which, if taken for the relation of involvement, would have the consequence that every true proposition 'involves' every other true proposition, and which would make true the thesis that there exists one and only one 'coherent' set of propositions. This relation is none other than the truth-functional '$p \rightarrow q$'. For according to the truth table defining it every true proposition implies every other true proposition. However, to define involvement by the truth-functional implication would be abhorrent to any coherence theorist, since he regards involvement as a relation between meanings and not truth-values. More seriously, the truth-functional conditional is itself defined in terms of truth and one could thus not without circularity define truth in terms of coherence, coherence in terms of involvement and involvement as the truth-functional conditional. (pp.106-7)

In fact there are **many** implication connectives ($\rightarrow$) which admit the rule: where A, B then A$\rightarrow$B, and so under unconstrained application have the (usually) undesirable result that any true proposition implies any other. And these connectives do not call for circular truth-functional characterisation. But **nor** does material implication ($\supset$). It is surprising that Körner should have missed the point that there are many ways of cooking a goose, all too many routes which lead back to material implication. One familiar way, among many, of characterising '$\supset$' is axiomatic; and however that is done, it will soon follow as a derived rule: where A, B then A$\supset$B.[7]

Let involvement be material implication, so characterised. Form a (indeed the) maximal consistent (nontrivial) set Γ closed under material implication, starting with any tautology or truth (e.g. Hegel's "Being is"). Then Γ is coherent, and uniquely so. And ⌜A⌝ is true iff A is in Γ. The coherence theory of truth is thus correct, and vindicated, if rather trivially so, given the massive historical backdrop.

It is true that some coherence theorists would likely find this way of fixing things up abhorrent, though perhaps less so than dedicated classical opponents of idealist metaphysics. But perhaps it would strike some coherence theorists as amusing, and even as Bradley's revenge. Certainly any idealist who finds classical logic the right medium in which to logically set his or her verificationist proclivities - as the logical positivists thought that this was the right medium for verification, with none of this new-fangled intuitionism - should not be displeased. As for us, well, we can have both correspondence and coherence along with Tarski-Bolzano theory duly generalised.[8] And no doubt other theories of truth can be worked into the synthesis too; but that's a task diverging from Körner and for another day.

3. Perennial Philosophy

What has been looked at, in some detail, takes off from a few pages in Körner's **Fundamental Questions**. There are many other pages that would repay similar concentrated study. But here there is room for only one final point, a corollary of what has been established. Körner claims, in the final paragraph of **Fundamental Questions** (p.280), that 'there is no perennial philosophy - no hard unchangeable core of philosophical truth'. Isn't there? Even in his own terms? Isn't the correspondence theory itself which is true, and which has held in virtually unchanged form from Aristotle to Körner, more than 2000 years, part of the hard core?

Körner says he has argued his claim, but it is not obvious where in his book any such argument occurred (certainly it is not cross-referenced and no track to follow is provided in the index). Charles Pigden, who made a detailed search for such an argument, without success, remarks:

> So far as I can make out Körner's argument for the "no perennial philosophy" thesis consists in his thumbnail history of philosophy. The philosophical agenda is set by the intellectual milieu - the total milieu - though there is mutual interaction. Thus the agenda of ancient philosophy is set by their budding cosmology, geometry and moral and political concerns, the agenda of medieval society by theology, and

modern philosophy by mathematics, science (and nowadays logic). Furthermore other disciplines split off from philosophy and became independent studies, etc.

But such an argument would not, on its own, guarantee Körner's claim. The undoubted differences in philosophy between different historical times and different cultures, does not show that there is not a common core. Compare the theme sometimes advanced that, though ethical systems vary substantially with different cultures, still there is common core of values and residual principles shared across cultures. Körner is particularly badly placed by virtue of his defence of various perennial philosophical themes as correct.

Furthermore there is, though Körner nowhere alludes to it, a philosophy called 'the perennial philosophy' (this description is used by Huxley and others). This philosophy **is** a perennial philosophy, not merely through its cheeky description, but by virtue of its key theses (which are unscrambled and derived in IAR, see footnote 29 there), which are both longstanding and also, so it can now be argued by semantical methods not unlike those used to establish the correspondence theory of truth, true.[9]

* * * * * * * * * *

FOOTNOTES

1. The biconditional 'iff' can be strong, e.g. of relevant strength given an appropriate background logic. A similar theorem can presumably be obtained in terms of other kinds of universal semantics, e.g. probabilistic, algebraic.

2. Most trivially by simply defining 'world', in one of its truncated senses, in the style of Wittgenstein, as the fusion of facts. But the larger endeavour, a fuller acount of 'world', will not be attempted here.

3. In even part-way decent logical theories, such as Russell's, which give some glimmer of intensionality, Davidson's argument to the Great Fact is fallacious: see Hochberg, p.279ff. In wider intensional settings such as universal semantics provide, the replacement principles essential to Davidson's argument are of course entirely unacceptable. For these reasons and others, Davidson's announcement of 'the failure of correspondence theories of truth based on the notion of fact'(p.49) is premature. Davidson's case is far from watertight; e.g. it depends critically on blockages imposed by a narrow objectual theory of quantification (p.48), which can however, as Davidson now concedes (p.xv), be avoided.

4. Though, interestingly, he appeals in criticism of Descartes to an intuitive, but non-standard, notion of logical consequence, under which the content of consequences is contained in that of the starting assumptions. Unfortunately for the criticism, this traditional containment notion can be explicated in at relevant framework (as is explained in FD). Indeed with a suitable non-logical implication and both necessary and contingent self-evident starting points, Descartes' program is not entirely dead, but can also be to some degree resuscitated.

5. To this small band the author belongs. And the project here, of making traditional theories of truth good, is really part of a more general enterprise (that of SMM) according to which, near enough, **every persistent metaphysical theory is correct**.

6. Curiously Körner insists on comparing involvement with **logical** implication, though he has indicated that it reflects a non-logical relation. Let us simply drop the qualification to logical.

7. There are further complexities in the background here, concerning completeness. For the present, let it be assumed that quantification theory or λ-categorial theory is axiomatised, and that every other theory is treated as a postulate theory in terms of one of these.

8. Tarski not only saw the semantic theory as complementing a correspondence theory but also was one of the first to explicate maximal consistent sets; he, if anyone, is behind this whole conspiracy.

9. My thanks to Jack Smart who fired my enthusiasm for the main topics addressed, through his persistent resistance to facts, to correspondence to the facts, and to coherence (see e.g. Smart).

* * * * * * * * * *

REFERENCES

Donald Davidson, **Inquiries into Truth and Interpretation** (Clarendon, Oxford, 1984).

H.Hochberg, **Logic, Ontology, and Language** (Philosophia Verlag, München, 1984).

S.Körner, **Fundamental Questions of Philosophy** (Harvester Press, Sussex, 1979).

V.Plumwood & R.Routley, "The Inadequacy of the Actual and the Real: Beyond Empiricism, Idealism and Mysticism", in **Language and Ontology**, (eds.) W.Leinfellner, E.Kramer & J.Schank (Holder-Pichter-Tempsky, Vienna, 1982), pp.49-67; referred to as IAR.

V.Plumwood & R.Routley, "The Semantics of First Degree Entailment", **Nous** 6 (1972) pp.335-59; referred to as FD.

R.Routley, "Universal Semantics?" **Journal of Philosophical Logic** 4 (1975) pp.327-56, also in adapted form in JB; referred to as US.

R.Routley, "The Semantical Metamorphosis of Metaphysics", **Australasian Journal of Philosophy** 54 (1976) pp.187-205; refered to as SMM.

R.Routley, **Exploring Meinong's Jungle and Beyond** (Research School of Social Sciences, Australian National University, 1980); referred to as JB.

J.J.C.Smart, "How to Turn the **Tractatus** Wittgenstein into (almost) Donald Davidson", **The Philosophy of Donald Davidson: A Perspective on Inquiries into Truth and Interpretation,** (ed.) E.LePore (Blackwell, Oxford, 1985).

* * * * * * * * * *

REPLY TO DR.SYLVAN

Stephan Körner

I found Dr.Routley-Sylvan's discussion of my views on theories of truth very valuable and welcome the opportunity to clarify some points. The following remarks are based on emphasizing certain distinctions which, I think, are also of independent philosophical interest. They are (1) the distinction between general concepts of truth, which cannot serve as effective criteria of truth, and specific concepts of truth which can be used in this way; (2) the distinction between inferential relations which are either necessary or ampliative and inferential relations which are assumed to be both, necessary and ampliative; (3) the distinction between philosophical theses which as a matter of empirical fact have so far been generally accepted by philosophers and philosophical theses which are assumed to be "necessarily true" because they can be and have been, demonstrated by some specific and allegedly wholly reliable philosophical method.

1. Correspondence and Facts

Routley-Sylvan shows that a theory of truth as correspondence or, indeed, isomorphism between statements and facts can be devised on the analogy of Tarski's theory of truth as correspondence between the statements of a metalanguage and the statements of the object-language of which it is the metalanguage. Thus, just as according to Tarksi's theory of truth the metastatement **aRb** is true iff **aRb**, so on Routley-Sylvan's "truth-fact" theory **aRb** is true iff it is a fact **that aRb**. The introduction of connectives, quantifiers and other operators proceeds by recursive definitions which in an equally obvious manner follow the recursive definitions of Tarski's theory.

If one agrees that a correspondence theory of truth, which relates true sentences to facts, can be constructed after the fashion of Tarski's theory, which relates true metastatements to object-statements, then one must also admit that the former theory shares the limitations of the later. Thus as Tarski points out, he did not attempt to provide a definition of 'true sentence' which would be **"in harmony with the laws of logic and the spirit of everyday language"** but a definition restricted to **"formalized languages"**.[1] Another limitation of Tarski's theory of truth and, hence, of Routley-Sylvan's theory is its being based on a version of classical logic – as opposed e.g. to intuitionist logic or a logic admitting inexact concepts. Tarski's and, hence, Routley-Sylvan's theory does not provide a definition of the general notion of truth, but only of a rather special one. At the same time the definition provides an effective criterion of truth for a limited region of thought, e.g. for the language and metalanguage of Boolean algebra and the corresponding region of thought, covered by Routley-Sylvan's theory.

The correspondence theory of truth which I discuss differs from Tarski's and, hence, from Routley-Sylvan's theory in two important respects: It provides, e.g. in its Aristotelian version, a general definition of truth and not one which is limited to certain regions of thinking. But it does not provide any criterion of truth. This is why Tarski's theory forms part of a special discipline, namely model theory or semantics, while the general correspondence theory belongs to transcendent metaphysics.[2]

2. On Coherence Theories

My main criticism of coherence theories concerns the allegedly logical relation in virtue of which all true propositions form a coherent system. This relation, which I called "involvement", has at least the following three characteristics: (a) If **p** involves **q**, then **q** can with necessity be inferred from **p**; (b) **p** and **q** is richer in content than **p**; (c) if **p** is a true proposition then it directly or indirectly (i.e. **via** the involvement of other propositions) involves all true propositions. Hegel's philosophy and more recently the philosophy of Brand Blanshard imply the thesis that all true propositions form a coherent system.

Insofar as involvement combines inferential necessity with the amplification of true information, it goes back at least to Descartes, who contrasts his notion of "deduction" which is necessary and ampliative with logical deduction which is necessary but not ampliative. It is easy to give examples of inferential relations which are necessary and not ampliative.

Among them are the Aristotelian syllogisms and the deducibility-relations of logicist and intuitionist logics. It is equally easy to give examples of inferential relations which are ampliative, but not necessary. Among them are the various relations of probability. But there are, as far as I can see, no necessary **and** ampliative inferential relations. That the Cartesian "deduction" is not necessary, can be shown by providing counterexamples, the acceptance of which does not lead to inconsistencies or some other breakdown of thinking. The same applies to other versions of necessary and ampliative involvement.[3]

In order to avoid misunderstandings, it will be useful to say a few words about material implications, which Routley-Sylvan regards as involvement-relations, and about Kant's synthetic **a priori** judgments, which Kant and others regard as necessary and ampliative. If we confuse material implication with logical implication then material implication will appear to be a species of involvement, since every true proposition materially implies every other true proposition and hence the conjunction of all true propositions. Yet a material implication, e.g. **'Brutus murdered Caesar & Paris is the capital of France'**, may be ampliative without being necessary (in the sense of Descartes, Hegel, Blanshard or any other coherence-theorist).

As regards Kant's synthetic **a priori** judgments, I should deny that they are necessary in the sense that without their acceptance thinking becomes impossible.[4] For, as I have argued elsewhere, the post-Kantian development of mathematics and of the natural sciences has shown that Kant's claim that the list of his synthetic **a priori** judgments is complete and unique is mistaken. Quite apart from this, Kant's epistemology is incompatible with any coherence theory of truth. For he does not accept the assumption of a necessary and ampliative inferential relation with the help of which every true proposition could be directly or indirectly inferred from any other, or a least, from a finite subset of true propositions.

3. On Two Conceptions of 'Perennial Philosophy'

Routley-Sylvan is right that in the last paragraph of my introductory book **Fundamental Questions of Philosophy** I do not make sufficiently clear in which sense of the term I reject "perennial philosophy". That I do so elsewhere[5] is, if not an excuse, at least an extenuating circumstance. I agree that if we compare the metaphysical beliefs and the moral attitudes of a wide variety of human beings, it makes good sense to distinguish between a common core, accepted by all of them, and different peripheries. I regard the acceptance of this common core as an anthropological fact and have tried to show that, and how, the ways in which human beings (as a matter of empirical fact) organize their beliefs and practical attitudes commit them **ipso facto** to the acceptance of a categorial framework, though not necessarily the same categorial framework, and to the acceptance of a morality, though not necesarily the same morality.[6]

I do, however, argue against the view that some human metaphysical beliefs or moral attitudes are necessary (and hence not merely unchanged but unchangeable) in a sense of "necessity" which can be demonstrated by some specific philosophical method such as the Cartesian method of doubt, the Kantian transcendental method, the Hegelian method, the "linguistic method" etc.. Indeed, just as I find it reasonable to assume that some of our evolutionary predecessors apprehended the world in ways which did not imply their having a metaphysics or morality, so do I find it conceivable that some of our evolutionary successors will apprehend the world in ways which also do not imply their having a metaphysics or morality, but which nevertheless may lead them to reflect on the manner of their apprehending the world, i.e. to pursue philosophy in an admittedly wide sense of the term.

* * * * * * * * * *

FOOTNOTES

1. See Sect.I of **the Concept of Truth in Formalized Languages** which appeared in Polish in 1931. The English version constitutes Ch.VIII of **Logic, Semantics, Metamathematics** (Oxford, 1956).

2. See **Metaphysics: Its Structure and Function** (Cambridge University Press, 1984) Ch.10.

3. Compare Sect.I of my reply to Professor Chisholm.

4. See e.g. **Kritik der reinen Vernunft**, B87.

5. E.g. in Chs.X & XV of **Metaphysics: Its Structure and Function** (Cambridge, 1984).

6. See **op.cit.** and **Experience and Conduct** (Cambridge, 1976).

* * * * * * * * * *

6. PRUDENCE AND AKRASIA

Robert A. Sharpe

1. Buried in the middle of Stephan Körner's discussion of prudence in **Experience and Conduct** is a rather uncharacteristic disclosure. We are invited to consider the possibility of generalising our prudential principles with respect to all persons who are self-regarding. What is prudent for me, if expressed in a sufficiently general manner, becomes a statement of the condition of everybody's happiness - "in a sense of the term which can be neither unreasonable nor immoral".

The extent to which this expresses an Aristotelian commitment to something like eudaimonia as the chief goal of man depends upon the centrality of Butlerian prudence in Körner's moral theory and he is chacteristically chary of committing himself to a substantive moral theory. His work in ethics has taken the form of trying to exhibit the logical structure of moral thought rather than that of defending any of the traditional alternatives such as Intuitionism, Kantianism or Utilitarianism. My own judgement is that the theory of prudence together with the theory of a hierarchy of attitudes upon which it depends, is much the most interesting, important and original contribution Körner has made in what might, in a generous sense of 'moral' be moral theory, though it more properly belongs to the adjacent area of practical reasoning.

I intend to examine his account of prudence, prefacing it with a discussion of the theory of attitude-stratification upon which it depends. The second and longer part of this essay is an analysis of akrasia for I believe that akrasia cannot be understood save as a form of counter-prudential behaviour. The conclusion uses this to point to what I believe to be certain flaws in Körner's conception of prudence. But first to the reportage.

1. It is perfectly possible for me to have inconsistent practical attitudes towards an object. I might want to have my cake and eat it but I

cannot, of course, do both.[1] Now I might have no attitudes at all towards these two alternatives. In Körner's technical terminology, neither of these first-level attitudes is dominated. On the other hand I may well wish to reduce my intake of sugar and starch. If this is expressed in the form of a first-level negative attitude, 'I do not want the cake', then, like any other Humean approach which takes all our attitudes as being of equal standing, it fails to match the phenomenology of the situation. The point is rather that I want the cake but know that I should not have it. In fact I think it imprudent to have it. Körner expresses this as my having a negative second level attitude to my first level attitude.[2] Now the possibility of having attitudes towards attitudes, being pleased that you have an attitude or deploring it, by no means exhausts the various interrelations attitudes have to one another. If it happens that an agent has a pair of attitudes like those in my example such that he can only act in accordance with one of them, the attitudes are described as practically complementary.[3] Furthermore it is possible, since we are not always consistent, to have what Körner calls logically incongruent attitudes; this occurs where we have positive second order attitudes to complementary first level attitudes. Thus suppose I have a positive attitude both to having my cake and eating it; these two attitudes will be complementary since I cannot do both. If on top of this I have a positive attitude to both first level attitudes, i.e. that I favour having a positive attitude to having my cake and favour having a positive attitude towards eating it, then I have two second level attitudes such that corresponding first level attitudes cannot both be satisfied. These he describes as incongruent. Moral attitudes, claims Körner, are, by definition, undominated and they are of at least second level. Furthermore they must be universalisable,[4] at least by the person who promulgates them. This attitude Körner describes as personally universal.[5] Also for an action to be obligatory not only the action that complies with it but the requiring of it must itself be moral.[6]

There are controversial matters here. Firstly too much philosophical water has passed under the bridge for us to be comfortable about an appeal to universalisability. But that calls for at least a book on its own count. I want to raise the lesser question as to whether all moral attitudes are necessarily undominated. There is one clear way in which this is false. Many people subscribe to a supreme moral principle which enshrines an attitude which is taken to dominate moral attitudes to concrete cases. The Utilitarian attitude to happiness or the Christian attitude to loving God are but two examples. It does not follow that my altruistic concerns for my neighbour cease to be moral because they are dominated by attitudes encompassed in these familiar overarching principles. Secondly it is pretty clear that we frequently ignore the stirrings of conscience in order to pursue our own pleasure. This may be weakness but it may also be a deliberate judgment that moral requirements of little weight, like keeping a minor promise made some time ago, should come second to matters of advantage to myself. I ought to see the student but the Vice Chancellor wishes to see me about an application for departmental funds immediately before the meeting. The latter is to my advantage. If I have a negative attitude to the second level attitude of wanting to keep promises on those occasions but have a positive first level attitude to securing a grant for research, does it follow that my attitude to promise keeping is not moral? Not a bit of it. As a matter of fact our moral attitudes are sometimes dominated by selfish considerations without losing their moral character. Indeed such a decision hardly seems irrational. Let us try a revision of Körner's Thesis. Suppose we have an obligation to place moral considerations above all others. Then I would argue in reply that even this modified version is implausible. It is not plausible to claim that moral consideration always ought to outweigh matters of personal advantage. Not only are moral attitudes in fact not dominating but we do not even have an obligation to have a system of attitudes such that the dominating attitudes are moral in character.

If we accept this then we are, I think, led away from a purely formal characterisation of what constitutes a moral attitude to a more substantial account in terms of the nature of the objects of these attitudes. For Körner makes universalisability and dominance defining conditions of the moral. As I have observed, I share most other philosophers' concern about the first and I have attempted here to produce counters to the second. Given this, then the purely formal account of moral attitudes given by Körner must fail.

Between the two extremes of undominated and universalised attitudes and the purely selfish attitudes which an individual might have to a concrete situation lie prudential attitudes. The form of prudence which interests

Körner is Butlerian prudence, that "reasonable self-love the end of which is our worldy interest".[7] This is further defined as dominated by a moral attitude. The point about prudence is that it is both generalisable and not immoral.[8] The generalising condition ensures that prudent behaviour is rational for any member of the class qua member of that class and distinguishes prudential from selfish attitudes. Actions which are moral but not prudent are acts of supererogation. The variety of relationships which we have already noted as existing between practical attitudes of various levels hold also in the case of prudence. There may be conflict between prudential attitudes. What is prudent for a Muslim fundamentalist who believes he goes straight to Paradise if killed in a holy war is not prudent for the reader of this book. But imagine that somebody is torn between life styles or brings relics of a former life style to his new existence. He may well have conflicting prudential attitudes as a consequence. We can define imprudence[9] as a matter of acting on first level attitudes which are contrary to those second level attitudes which express the prudential outlook. Note that the consequence of treating prudence as normally dominated is that though immoral action may be either imprudent or indifferent it cannot be prudent.

In order to develop an objection to Körner's notion of prudence I shall assume a connection between attitude and desire. The connection is required in any case by Körner to make the link between practical reasoning and action. It is desires and intentions which translate practical attitudes into actions. With this additional premise I propose to develop Körner's theory of the stratification of action into an account of akrasia and, what I believe to be a neighbour of akrasia, self-deception. The more acute disagreements with Körner come over the scope of irrationality.

2. What is the problem about akrasia? It is obvious, is it not, that sometimes we act unwisely and against our own interests and fail to do what we should. We are weak willed. A paradigm case is that of the man who ought not to drink because he is driving home and yet, liking a drink, succumbs to the temptation. We say that if his will was stronger he would have acted differently. We often think of weakness of will as occuring when we suddenly succumb, but we can succumb, if not with deliberation, certainly rather slowly and predictably. In **Love in a Cold Climate,** the fop Cedric

predicts his own downfall:

> As I got back to my own house the telephone bell was pealing away - Cedric.
> 'I thought I'd better warn you my darling that we are chucking poor Norma tonight.'
> 'Cedric, you simply can't, I never heard anything so awful, she has bought cream!'
> He gave an unkind laugh and said:
> 'So much the better for those weedy tots I see creeping about the house.'
> 'But why should you chuck, are you ill?
> 'Not the least bit ill thank you love. The thing is that Merlin wants us to go over there for dinner, he has got fresh **foie gras,** and a fascinating Marquesa with eyelashes two inches long, he measured. Do you see how One can't resist it?'
> 'One must resist it,' I said, frantically. 'You simply cannot chuck poor Norma now, you'll never know the trouble she's taken. Besides, do think of us, you miserable boy, we can't chuck, only think of the dismal evening we shall have without you.'
> 'I know, poor you, won't it be lugubrious?'
> 'Cedric, all I can say is you are a sewer.'
> 'Yes, darling, **mea culpa.** But it's not so much that I want to chuck as that I absolutely know I shall. I don't even intend to, I fully intend not to, it is something in my body will make me. When I've rung off from speaking to you, I know that my hand will creep back to the receiver again of its own accord, and I shall hear my voice, but quite against my will, mind you, asking for Norma's number, and then I shall be really horrified to hear it breaking this dreadful news to Norma.'

Remember too, Austin's delightful account of dinner at high table.[10] The bombe is divided into equal parts corresponding to the number of diners. I fancy two slices of ice cream and take them. "Do I raven, do I snatch the morsels from the dish and then wolf them down impervious to the consternation of my colleagues? Not a bit of it." (A cartoon would be appropriate).

I shall follow current philosophical usage in my use of 'akrasia'. It is more frequently translated by 'incontinence' than by 'weakness of will', though both Davidson and Taylor[11] take it to be aptly rendered by the latter. The problems of finding an equivalent in English are highlighted by this uncertainty, for we do not think of somebody who cannot control his bowels as being weak willed, or at least not normally. Pears follows Liddell and Scott in recommending 'lack of control'.[12] The problem about akrasia is not dissimilar to the problem about self-deception. The problem with self-deception is that it invites construction on the model of lying. Self-deception is then lying to oneself. But there is evidently something quite extraordinary about the notion of deceiver and deceived being the same person. If you are in the position of deceiver, you can hardly be deceived. You know what is going on because you are doing it. Equally, there is a tendency to suppose that it cannot be the case that you do not want to do what you do as long as you act voluntarily. After all, if you are a free agent, what you do is at your own behest. You are the author of your acts and it can neither be the case that what you are doing is what you would rather not do, or that you wish you were doing what you do not do. And yet we do not think Saint Paul was guilty of some sort of logical error when he said "That which I would, I do not and that which I would not, that I do."

The crudest form of evasion of the problem follows Freud in distinguishing different agents within the same person. But this route immediately brings us up against further difficulties. Neither akrasia nor self-deception can occur unless the same individual is the subject of whatever is going on. The deception is a deceiving of oneself. Akrasia is a failure to do what one recognises is the best. Not much is to be gained by replicating the logical problems of the doctrine of the Trinity.

However, if we, alternatively, stress the unity of the agent, we may equally bring about the collapse of akrasia into something else. First of all we may think of the agent as being overwhelmed by his desires. This is the ever-present threat for a philosopher who construes desires as causes, taking action to be determined by the preponderance of casual factors. The agent desires to have the drink and also not to drive home under the influence. The desire to have the drink simply outweighs the desire not to. But the internal unity of the person in the homogeneity of factors determining agency is bought at a price and the price is that akrasia disappears. It is now no more than one form of conflict between desires in which the strongest desire triumphs. It is at this point that the virtues of a stratified theory like Körners become apparent. If we take a pro-attitude to x to entail a desire that x be realised and if we also allow that we can have positive or negative attitudes to attitudes of first level then we have the technical resources required. We can now allow that we may have a first level desire without having a second level desire that it be satisfied.

The second way in which I described how akrasia collapses has a variant. If we take desires to be reasons for action and take the totality of desires in this case as the totality of reasons, then we can conclude that the agent both has a reason for taking the drink and a reason for not taking it. In such a case such a decision to drink represents the way the reasons fall, the way the weighting turns out. Providing no miscalculation has taken place, the decision to have the drink is not irrational after all. Such a conclusion denies one of the main characteristics of akrasia, and that is that the akrates is irrational. For an irrational act is taken as being against the reasons and the taking of the drink cannot be said to be against the reasons if the reasons in totality favour it.

The especially problematic form which akrasia takes for Davidson[13] derives from his regarding desires as on all fours with reasons in precisely this way, for he thinks of as desires as reasons for action. It then became incumbent upon him to find a way in which akrasia can occur within this context. What he maintains is that the action which is akratic is, all things considered, not the best but that, relativised to what the agent desires, it is grounded in those desires. But this makes akrasia what it is not, a species of intellectual failure, for the agent has failed to take into consideration all the relevant factors. It comes close to the Aristotelian conception of akrasia as a temporary lapse of knowledge. In this case some of the factors which are pertinent to the decision have simply not been taken into account. It is also, I believe, a fundamental error to suppose that to have a desire for x is necessarily to have a reason for securing x. It may not be so. To have a reason for x-ing or for getting x requires that x either be seen as worthy in itself or as a means to a worthy end. Neither is necessarily fulfilled in the case of a desire. **Pari passu**, if I have a pro-attitude towards something it by no means follows that I have a reason for securing it.

The traditional rejoinder at this point is to argue that getting what you desire gives pleasure and that the pleasure is the good that confers reason status. Certainly, if you argue to yourself that if you satisfy your desire or match your positive attitude with action you will get pleasure and that the pleasure is good, then you have a reason to secure x or to do x. But it is bad philosophical psychology to claim that this is universally the case. As Bradley[14] pointed out at length, our actions are not often motivated by the belief that pleasure will follow. Normally I do x if I desire to do x, simply because I desire to do x. Of course, once ratiocination about pleasure enters the picture we have the possibility of a conflict of reasons. Suppose the balance of reasons favours not doing x but that I still do x, do we have akrasia? The answer is that if the agent draws the wrong conclusion from the balance of reasons then we have intellectual failure. If that intellectual failure is itself motivated there is, as I shall argue, a further form of irrationality. However if I do x because I desire to and take no notice of the result of the ratiocination, then we have the commoner form of akrasia. In this case the reasoning is no more than conceptual icing upon a volitional cake. The point upon which I want to insist is that characteristically when I do x I do it because I want to and that I neither do it nor want to do it because I reflect upon the pleasure I will get. Some actions, after all, are the results of whim and these are only the most obvious counter examples. Although desire mediated through reflection on pleasure does occur, I think it is uncommon. In any case the pleasure itself may not be good: the pleasures of gloating or of sadism cannot count as reasons for doing x unless the agent's value system is also perverted. But the case has been, I think, sufficiently made out without our needing to consider such exceptional circumstances.

A major flaw in Davidson's analysis of akrasia is that in his principle P2 he subscribes to the implausible thesis that if A has a reason to do x then he wants to do x.[15] (I take 'it would be better to do x' entails that he has a reason to do x). The principle reads as follows: "If an agent judges that it would be better to do x than to do y then he wants to do x more than he wants to do y." I assume that the non-comparative form would also be acceptable to Davidson. Conversely Davidson believes that the converse holds and that to have a desire is to have a reason. Following Aristotle he describes the desire to know the time as being conceived as a principle of action whose natural propositional expression would be something like 'It would be good for me to know the time.'

There is a third way in which akrasia may disappear. The agent may revalue his position. He looks at the situation and decides that there are no strong reasons against taking the drink. The difference is that here we no longer think of the agent as simply the plaything of his desires and his action as determined by the preponderance of factors bearing upon him. Rather we think of him as actively weighing the pros and cons. He is an agent. Davidson prefers this view of the will as a **tertium quid.**

If this is conceded however, there seems no good reason why Davidson should not concede further that we may adopt attitudes towards our desires; for if the agent weighs the pros and cons he does so on some basis and the obvious basis is that he has attitudes towards them. Having conceded this much there now seems no reason why we should any longer think that wanting to do x either implies or is implied by thinking that it is better to do x.

The Socratic problem about akrasia depends upon akrasia collapsing into one or other of the last two forms. In both cases we do what seems to be good, in the first case because the reasons turn out that way (desires count as reasons), and in the second because we have revalued. The puzzle then was how akrasia could occur at all? How could a man see the good and not desire it?

It has been claimed that the revaluation could be intellectual failure. Either I am mistaken about the goodness of the drink, or in the derivation of the conclusion of the practical syllogism there is some fairly gross error. I want to suggest that there are strong grounds for doubting whether I can innocently mistake the grounds for my action or innocently draw wrong conclusions from then. We are asked to believe that my judgment that I can take the extra drink could be just a mistake; I somehow miscalculate the force of the grounds against having that extra glass. But is this plausible? Most of us would, I think, be inclined to conclude that anybody who reasoned thus was self-deceiving.

If such a revaluation is innocent then the indications of self-deception do not correspond to the facts. Let us try to imagine a case! It might be that he is 'saying in his heart' that the reasons are thus and thus. Suppose he says to himself 'I am only taking this drink in order to mislead the onlookers into thinking that I am weak willed', or 'I am only taking this way in order to suggest that I am self-deceived'. We could suppose that he thinks it in his long term interests to give an impression that he is not fully in control. Perhaps he is about to clinch an important business deal, for example. Does this private performance constitute an internal fact about him with which the behavioural criteria clash?

The problem is, of course, whether we (or he) should take what he says 'in his heart' at face value. He can simply compound the self-deception by telling himself a story and once again the check on the truth of that story comes from how it meshes with grosser behavioural features. He, of course, cannot check on this for no accurate check could leave him self-deceived, although the checking could involve deception. Either he is not self-deceived or he discovers that he is self-deceived and hence is no longer. (You cannot know that you **are** self-deceived but only that you **were** self-deceived.) However, if he tells us his private thoughts, we could judge that he is self-deceived and we judge in the usual way. We watch the alacrity with which he rushes to the bar etc. What I suggest, then, is that the grounds must always be behavioural and there is no level at which the facts could **really** be such as to cancel **that**.

It is a mistake to suppose that there is an internal state of affairs, a heart of darkness of the mind, with which judgment about motives, about emotions or self-deception, either agrees or fails to agree. I cannot make out a full case for this here, of course, but the nub of this observation is that certain forms of realism about the mental are untenable.

Akrasia typically, though not necessarily, occurs where a long term commitment clashes with a short term desire and such long term desires we think of as prudential. The agent in my example wants to have the drink but also wants to be a responsible human being, and not somebody who has run down some innocent bystander under the influence of drink. 'Want' cannot be replaced by 'desire' with perfect propriety here. For he does not desire to be responsible.[16] 'Desire' is appropriate for wanting food, sex, material gew-gaws etc. The celibate may have desires which he does not **want** to gratify. Note that this difference mirrors Körner's distinction between first and second level attitudes. We can say quite properly that he has a reason for not driving home, whereas we cannot say that he has a reason for taking drink (namely, that he wants to). And, as we have seen, he might not want to accede to the desire. He has a negative attitude to the first level desire. I have a reason to do x only if x is either good in itself or a means to a good. So a clash between reasons and desire is always possible: I may have a reason not to do x because it is bad in itself or because it has bad consequences (or may have bad consequences), and yet I may desire x. So though I desire to have the drink, if I am strong willed I will not take the drink, assuming the reason is known to me.

PRUDENCE AND AKRASIA

It is natural to think of desires as relatively short lived and reasons as essentially long term. So it is worth pointing out that sometimes I can forget the reason I have for abstaining from x whilst the desire for x remains as strong; the reason still applies, of course, but it is no longer part of my ratiocinative stock, so to speak. Equally, my desire to improve my piano playing can be a long term desire, staying with me throughout my life, whilst the reason I have to offer first aid to the accident victim only exists as long as he is injured and nobody else helps him.

These cases are, however, somewhat exceptional. As Kenny[17] observes, general commitments show themselves over a lifetime. Although the primitive sign of wanting may be trying to get,[18] nevertheless this cannot be translated into a simplistic rule that unless you try to get x whenever possible, then you do not want x.

My thesis, then, is that we cannot understand the nature of akrasia unless we recognise that desires, like attitudes, are stratified and that prudence is a pivotal concept in these contexts. For akratic behaviour is counter-prudential. On the face of it, then, Körner's account displays exactly the features required to elucidate this most intractable puzzle within practical reasoning. **En passant,** it may be as well to point out that it is the agent's estimate of what is required to meet his long term ends which counts in assessing his behaviour as prudent. That he fails to see that this behaviour is, as we say, counter-productive or that it will not produce the intended effect does not detract from the fact he acts prudently. Of course he may be imprudent in his assessment of what is technically required to secure the end in view; he may be careless or over hasty in making up his mind but that imprudence has, so to speak, a different location.

The flaw which, in my view undermines his general account of prudence explains perhaps why Körner gives no general account of akrasia. As we have seen, Körner takes prudence to be morally dominated. Not all prudent behaviour is moral: it may be moral but it can also be morally indifferent. But it cannot be immoral. Behaviour to which I properly take a negative moral attitude of second level or higher will not be prudent behaviour. Thus Körner cannot say that the burglar was prudent in concealing his getaway car in the locked up garage near the bank nor can he say that he was imprudent in drinking so heavily before the raid. This strikes me as counter intuitive. I think prudence and imprudence are technical notions relativised to the aims of the agent.

Dorothy Walsh[19] distinguishes two forms of akrasia, moral transgression and prudential folly. Commonly, of course examples are of the latter. The man who takes the extra drink is immoral certainly, but it is not because of that that we regard him as akratic; he is akratic because he is imprudent and knowingly imprudent. You risk misery and the loss of much which you value if you drink and drive. Or take the case where, despite your commitment to a low cholesterol diet, you succumb to temptation and take the extra slice. The reasons which conflict with desire in akrasia are prudential. Thus a moral conflict where desire wins over duty does not count as weakness of will even though prudential elements may be present where the motivation is complex. I ought to visit a bereaved friend but Manchester United are playing Liverpool on television and I succumb to the temptation to stay home and watch the match. I act selfishly but I do not display weakness of will.

This conclusion is reinforced by two arguments. Firstly, all weakness of will is irrational. But the neglect of duty in favour of desire is not contrary to our current conception of rationality and hence is not weakness of will. Secondly, consider the scope of objects of akrasia. Prudential considerations are concerned with my own future, of course. But I can act akratically with respect, and only with respect, to those individuals whose interests I cannot think of as separate from my own. If my small child asks for sweets, I might indulgently give him sweets. Indulgence is like self-indulgence and self-indulgence is often akrasia. I may know that I have a reason not to give the child sweets. It is imprudent because it is bad for his teeth. We can extend this to any individual to whom I am, metaphorically or actually, in **loco parentis**. To give somebody else's children sweets may be akratic, though it is perhaps moral weakness and not akrasia to encourage imprudence in an autonomous adult. The point is beautifully made by Aristotle. "Parents love their children as themselves, for having been split off from them they are, as it were, other selves, selves at a distance." So prudence and my interest, intermesh in the way my thesis predicts.[20]

To summarise Körner's account:

    1. Actions may be moral and prudent.

2. Actions may be moral and imprudent. e.g. acts of supererogation.

3. Actions may not be immoral and prudent.

A question remains; is the following an entailment?

4. 'If an action is immoral, then it is imprudent.'

Initially this seems unlikely since the prudent/imprudent opposition does not exhaust the class of actions. An action like crossing my legs or scratching my ear is neither prudent nor imprudent. These actions have no long term effects of any significance. So an action might be immoral and fall into neither category. But it may well be argued that the category of moral actions is essentially that of actions which do have significant long term or at least medium term consequences. They are not actions of whim or actions of no moment. They are precisely those kinds of action where questions about prudence can be sensibly raised. If this is so then any action which may be judged moral or immoral can also be judged to be prudent or imprudent. The implication is that all immoral actions must be imprudent (since they cannot be prudent) and this brings us close to the substantial moral theory which I have argued Körner requires and yet to which he seems reluctant to commit himself. It is, as I predicted, Aristotelian in form.

It does not follow from this that all immoral actions are akratic. I only argue that all akratic actions are imprudent, not the converse. We may after all negligently fail to consider the consequences of an act and an act done in haste or without due consideration will be imprudent without being akratic. If the negligence suggests self-deception we may suspect irrationality. Such irrationality is weakness of will. The self-deceiver is weak willed in that a desire that the facts should be one way rather than another clashes with his long term commitment to making his belief accord with facts. For example a man may not want it to be the case that his son is an embezzler and turns away from the evidence which suggests that conclusion. In general he has a positive attitude of second level that his beliefs be true beliefs; in the particular case this requires that he faces the evidence squarely and sees where it leads. This he fails to do and his failure is a weakness of will. The important thing here is that it shows itself in his actions. He fails to investigate where he should. It is not like wishful thinking where there is just no evidence to support the wished-for conclusion. The evidence runs counter to the beliefs of the self-deceiver and, typically, he simply does not pursue the evidence to see where it leads. Obviously there are cases where the evidence forces itself upon my attention, where I have no alternative to believing what I believe. There might seem to be a parallel here with those cases where it is perfectly obvious what I ought to do if I am to obtain some particular end. But there are differences. I always can choose not to do something whilst it is in my power, whereas I cannot refrain from believing what I recognise the evidence supports. It is for this reason that I choose to speak of the self-deceiver as displaying his irrationality in the gathering of evidence. It is what he fails to do rather than what he fails to believe that shows his self-deception.

Readers may have noticed that I have not described the irrationality of the self-deceiver as akratic. Like the akrates the irrationality of the self-deceiver lies in his action but the self-deceiver, though weak willed, is not akratic. Classical accounts of the akrates show him as losing self-control or succumbing to the weight of desires or passions, or as being unable to resist an urge. We cannot think of the failure to investigate the facts in quite these terms. Indeed the attempt to construct a parallel shows that in certain respects the two cases are quite opposite to one another. In order to think of the failure to investigate as akratic we must think of ourselves as having a passion or urge not to investigate which somehow overwhelms the desire to find out how the facts lie. That is implausible. The likely case is in fact quite opposite; we could think of a person being possessed by curiosity which he is unable to restrain. That does sound like akrasia. But omissions do not.

Davidson's classic presentation of the problem of akrasia involves three elements:[21]

P1 If an agent wants to do x more than he wants to y and he believes himself free to do either x or y then he will intentionally do x if he does either x or y intentionally.

P2 If an agent judges that it would be better to do x than to do y, then he wants to do x more than he wants to do y.

P3 is that there are incontinent actions and this,

with P1 and P2, forms an inconsistent triad.

Let us allow that in deciding whether to believe p or not p we need to find out whether p or not p is true. So a parallel to Davidson's principles P1 and P2 will be:

> P1' If an agent wants to find out whether p is true more than he wants not to find out and he believes himself free to either find out or not find out, then he will intentionally find out if he either finds out or does not find out intentionally (assuming the truth of p is discoverable).
>
> P2' If an agent judges that it would be better to find out what the case is than not, then he wants to find out if p more than he wants not to find out.

Now obviously P2' is as weak as P2. From the fact that I judge it better to find out it does not follow that I want to find out. The self-deceiver, I propose, is precisely in this position. He judges it better, as a matter of general prudence, that his beliefs match the facts. Just as it is imprudent to drive under the influence since I risk getting caught, even though the chances may not be high, so it is imprudent not to match belief and fact, even though the consequences of not doing so might not be particularly damaging in this case. But to match belief and fact commonly requires finding out what the evidence is, and this is a matter of action. Characteristically the self-deceiver has an inkling that the evidence might prove damaging to cherished beliefs, and he does not pursue it. The irrationality of this is not the stark irrationality of self-contradiction but the irrationality of akrasia, which is in turn the irrationality of imprudence. Indeed, without the stress on imprudence, it is hard to see what is wrong with self-deception. It is not obviously immoral.

We can easily imagine a woman not wishing to investigate her lover's behaviour further. If she finds out that he is unfaithful the consequences will be so enormous in terms of her self-esteem and the patterns and structure of her life that it is, it might be claimed, more prudent to be self-deceiving. How can I reply? It is understandable that she should so react but I believe that it is more prudent to build one's life on sure foundations. Although I think that the estimate of what is prudent is much harder where self-deception is an option than with straightforward akrasia

I still think that in the long term human relationships are safer when built on honesty. After all, the suspicions have already done their work. There may be reasons against investigating, since investigations already indicate a lack of trust which soils that relationship; but ultimately those who counsel knowledge do so for prudential reasons. When we give this sort of advice we should do so in fear and trembling and it is precisely because of this damage of trust that self-deception seems less irrational than my initially rather blunt presentation might have suggested. Yet if there are features which ameliorate its irrationality they are compensated for by the conflict which I described between our general attitude that truth seeking is good and our special pleading on behalf of our own beliefs. That conflict, like the special favours we give to our religious or ideological beliefs, is an inconsistency. Santayana put it well: "However unpleasant truth may prove, we long to know it, partly because experience has shown us the prudence of this kind of intellectual courage, and chiefly because the consciousness of ignorance and the dread of the unknown is more tormenting than any possible discovery." None of these contingencies, of course, affects my central conceptual claim that self-deception is a failure in action, a failure to investigate.

So to the objection that there is something strange about calling self-deception imprudent, I reply that there is nothing strange about calling a failure to investigate imprudent, and it is just there that the nub of deception is to be found. The enormous advantage of this account of self-deception is that it accounts for the phenomenon without any special difficulties connected with 'willing oneself to believe'.

3. What I have argued is that a Körner-style stratification of attitudes provides us with the ontology we require to understand the problem of akrasia and self-deception. Where I disagree with Körner is in his thesis that prudence is a morally dominated concept such that it cannot be said of an immoral man that his behaviour is prudent. Prudence is a matter of acting consistently with our conception of our interests and morality is irrelevant to it. On his conception of prudence, Körner comes close to an Aristotelian conception of the good for man, namely that living well is prudent for by doing what is right we do the best for ourselves. Thus even if, in the short term, it seems to me to be in my interests to increase my

wealth at the expense of others in the long run this unethical conduct damages me as well.

One can imagine, of course, a philosopher maintaining this as a substantive moral position. 'I believe that immorality is imprudent.' But this would be a matter of replacement rather than exhibition analysis and comes close to endorsing what has become known as 'High-minded egoism'. I suspect that most of us would have found Aristotle a bit of a prig.[22]

* * * * * * * * * *

FOOTNOTES

1. **Experience and Conduct** (Cambridge University Press, 1976), p.91.

2. **ibid**, p.93.

3. **ibid**, p.96.

4. **ibid**, p.139.

5. **ibid**, p.138, p.142.

6. **ibid**, p.144.

7. **ibid**, p.163.

8. **ibid**, p.164.

9. **ibid**, p.171.

10. J.L.Austin, "A Plea for Excuses", **Philosophical Papers**, (Oxford University Press, 1979), p.198.

11. D.Davidson, "How is Weakness of Will Possible?" **Essays on Actions and Events** (Oxford University Press, 1980); C.C.W.Taylor, "Plato, Hare and Davidson on Akrasia", <u>Mind</u> Vol.89 (1980).

12. David Pears, **Motivated Irrationality** (Oxford University Press, 1984) p.23.

13. Davidson **ibid**, p.31, also pp.3-4, 15.

14. **Ethical Studies.**

15. Davidson, **ibid**, p.23.

16. See Taylor **ibid**, for an excellent discussion of the complexities of desire.

17. A.Kenny, **Aristotle's Theory of the Will** (Duckworth, 1979) pp.165-6.

18. G.E.M.Anscombe, **Intention** (Blackwell, 1957) p.68.

19. **Ethics**, 1975.

20. Aristotle, **Ethics** VIII. 1161b.

21. **ibid**, p.23.

22. I am indebted to Dr.David Cockburn for his comments on this paper and to Dr.Julius Tomin who corrected some misapprehensions about Aristotle. I have also been helped by the discussions of the relationship between self-deception and akrasia by David Pears in **Motivated Irrationality** and by David Charles in "Rationality and Irrationality", **Proceedings of the Aristotelian Society** 1982-3.

* * * * * * * * * *

REPLY TO PROFESSOR SHARPE

Stephan Körner

It gives me much satisfaction that Professor Sharpe on the whole accepts those features of my account of practical thinking which seem to me central to the understanding of its structure and function, even though he disagrees with me on some issues which seem to me less clear-cut than they seem to him. He agrees with what I have to say about some of the global aspects of practical thinking, in particular about the stratification of practical attitudes and about practical, as opposed to theoretical inconsistency; and he regards these notions as useful tools for the analysis of specific concepts. Yet when it comes to the use of these tools, he sometimes disagrees with my results, in particular my analysis of moral principles as supreme and my analysis of prudence as morally dominated. Part 1 of the following remarks deals with the problem of the supremacy of moral principles. Part 2 contains a defence of my conception of prudence, which is morally dominated, while admitting the non-emptiness of a different conception of prudence which is not. It also contains some comments on Sharpe's analysis of akrasia.

1. Before considering Sharpe's criticism of my definition of moral attitudes as being at least of second level, as being undominated and as universalizable, his formulation of the definition must be corrected. He rightly points out that the problem of universalizability is very complex, but fails to mention that my definition requires universalizability in the sense in which the practicability which is the object of the second level practical attitude (or more generally, the highest level practical attitude) is a state of affairs in which **everybody** has a certain practical attitude. It also seems to me that he does not take sufficient notice of the distinction between practical and other attitudes. Thus a person's **practical** pro-attitude towards a practicability (a state of affairs which he rightly considers realizable) involves his determination to bring it

about, if its realization depends only on himself, or to contribute to its realization if it also depends on others. A person's practical anti-attitude towards a practicability similarly involves his determination to prevent its realization or to contribute to its prevention. A person's practical attitude of indifference towards a practicability involves its acknowledgement as a practicability to the realization or non-realization of which he may or may not contribute. Practical attitudes differ from pure attitudes, which are directed towards states of affairs, independently of their being realized, unrealized, realizable or unrealizable. This is not the place to consider the variety of pure attitudes, e.g. aesthetic attitudes.[1] I may have a pure pro-attitude not only towards being eternally young, but also towards the practicability of looking as young as possible - if I am not determined to try bringing it about.

In my analysis of practical thinking I am concerned with the stratification of **practical** attitudes. That is to say that a practical attitude is (practically) undominated if, and only if, it is not the object of a higher **practical** attitude. A practical attitude does not cease to be practically undominated or supreme by being also the object of a pure attitude, e.g. an aesthetic attitude. It seems useful to give a schematic example of a moral principle.[2] A suitable example is the following expression of a moral pro-attitude of second level:

$$S + /U+X/$$

in words "The person S has a practical pro-attitude towards a state of affairs in which everybody including himself has a practical pro-attitude towards the realization of a state of affairs characterized by X". (What will be said about moral attitudes of this form applies with obvious modifications to moral attitudes in which the place '+' is taken by '-' or '±', which respectively symbolize a practical anti-attitude or attitude of indifference. It also applies to moral principles of higher level.)

It is now possible to consider two objections which Sharpe directs towards my thesis that moral attitudes are supreme, i.e. not dominated by higher practical attitudes. One objection is that "many people subscribe to a supreme moral principle which enshrines an attitude which is taken to dominate moral attitudes to concrete cases". Examples of such enshrinement are "the Utilitarian attitude to happiness or the Christian attitude to loving God". Neither example seems to me to affect my thesis. As regards the first, it seems sufficient to point out that utilitarian morality fits well into my schema: A person is a utilitarian if, and only if, he has an undominated practical pro-attitude towards a practicable state of affairs in which everybody including himself has a practical pro-attitude towards the realization of a state of affairs in which the greatest realizable happiness of the greatest number of human beings is being realized. (This rather crude version of the utilitarian criterion of morality may, of course, be replaced by a more refined one). - As regards the Christian attitude to loving God, it does not practically dominate a Christian's morality. Being a Christian implies accepting the Christian morality as having been laid down by God, as well as loving God. But the latter pro-attitude is not a practical attitude, even though it may help the Christian in conforming to the Christian morality. (See **E&C** pp.200f.)

Sharpe's other objection to the thesis that moral attitudes are supreme is based on situations in which one makes "a deliberate judgment that moral requirements of little weight, like keeping a minor promise made some time ago should come second to matters of advantage to oneself". As I see it, the deliberate judgment which excuses the non-fulfilment of the minor promise in certain circumstances is a moral judgment. The position is similar to legal thinking in which an appeal is made to the principle that **de minimis non curat lex.** Just as the appeal to this principle is an appeal to an accepted legal principle, so is the appeal to an overriding personal interest in certain more or less precisely given circumstances an appeal to an accepted moral principle. That the decision about the role of the personal interest is a moral decision may be overlooked for a variety of reasons. Among them are the mistaken assumption that a moral principle may not admit of exceptions, the conflation of moral rules of thumb with strict moral principles and the assumption that an apparent conflict between competing moral principles can be solved by an appeal to a non-moral attitude. (See **E&C** Chs.9-12).

2. Sharpe objects to my account of Butlerian prudence because I define it as not-immoral, i.e. as (positively or indifferently) dominated by a moral attitude and, of course, as not otherwise inconsistent with any moral principle. For the sake of brevity it is useful to compare the schematic formulation of a prudential principle in my sense with the earlier sche-

matic formulation of a moral principle:

$$S + /U+X/$$

A prudential principle which exemplifies my definition may have the form:

$$...S + /G+Y/$$

The dots in front of S refer to the principle's being dominated by a pro- or indifferent moral attitude of S. S's prudential pro-attitude is directed towards a practicability in which every member of a group G to which S belongs has a practical pro-attitude towards a practicability characterized by Y. To give an example, as a teacher of philosophy for many years (as a member of a group G) I have a practical pro-attitude towards a practicability which consists in every philosophy teacher's having a practical pro-attitude towards giving lively lectures. My practical attitude is to some extent self-regarding, e.g. in trying to avoid the displeasure of looking at yawning students. It is personally general, but not universal. And it is dominated by a positive moral attitude, namely the moral attitude towards everybody's fulfilling his professional tasks to the best of his ability.

I do not think that Sharpe would doubt the sincerity of my statement that the practical attitude which I have just described and which is dominated by a moral attitude and in general required to be consistent with my morality, is as a matter of empirical fact one of my practical attitudes. What he objects to is my calling this practical attitude "prudential". My answer to this objection is conciliatory: While my reading of Butler and my feeling for the English language may have caused me to misname the concept which I called "prudence" and which I accept in my practical thinking, I do think that I am not alone in using this concept and that it is proper to characterize it as a useful and frequently used concept of practical thinking. Moreover, since Sharpe's objection only concerns my definition of prudence as morally dominated, I might - without throwing any doubt on his scholarship or introspection - propose to distinguish morally dominated (Butlerian, morally constrained, at least not immoral) prudence on the one hand and morally undominated (morally unconstrained) prudence or efficiency on the other. Sharpe holds that on my view any immoral action is imprudent. But this is surely not so, since my performing an immoral action such as committing a murder or a fraud would have nothing to do with my observing the prudential principles which I have accepted as a teacher of philosophy.[3]

Sharpe's remarks on akrasia are meant to support what he has to say about the notion of prudence. For Sharpe akratic behaviour is counter-prudential in a sense of the term in which prudence is not morally constrained. His notion of akrasia differs from Aristotle's who says of the akratic man that "he resembles a state which passes all the proper laws, but never keeps them."[4] It is not difficult to do justice to the variety of conduct which philosophers and others have called "akratic", "incontinent" or "weakwilled". Thus we may distinguish between akrasia in Aristotle's, Butler's (or, to be more careful, my) and Sharpe's sense. The Aristotelian akratic may, for example, accept a moral princple of form

$$S + /U+X/$$

and yet perform an action characterized by **non-X**. The Butlerian akratic may, for example, accept a morally dominated prudential principle of form

$$... S + /G+Y/$$

and yet perform an action characterized by **non-Y**. Last, the Sharpean akratic may, for example, accept a prudential principle of form

$$S + /G+Z/$$

which is to be considered independently of the agent's morality, and yet perform an action characterized by non-Z. What I have to say about Sharpe's subtle discussion of self-deception is **mutatis mutandis** very similar to what I have said about his discussion of prudence and akrasia.

In conclusion I should like to emphasize once again that my main philosophical interest in practical thinking is not to defend my own moral convictions, but to analyse the global structure of practical thinking. It is in particular to characterize the **genus** morality and, thereby, to provide an instrument for comparing its **species**. In applying an instrument for its intended task, one may of course come to the conclusion that the instrument stands in need of some refinement. (See my reply to Professor Marciszewski). It is by no means impossible that further reflection on Sharpe's essay may convince me of such a need.

* * * * * * * * *

FOOTNOTES

1. See e.g. **Metaphysics: Its Structure and Function** (Cambridge University Press, 1984) Ch.3.

2. For details see **Experience and Conduct** (Cambridge University Press, 1976) Chs.9-11, henceforth referred to as **E&C**.

3. See **E&C** p.164, where I admit that 'prudence' may be a misnomer for the concept which I discuss under that name.

4. **Nicomachean Ethics** VI, X, 3.

* * * * * * * * *

7. DETERMINISM, RESPONSIBILITY AND COMPUTERS

Jacek Hołówka

In his **Abstraction in Science and Morals** Stephan Körner has provided a very useful framework for an assessment of the view that human beings are capable of making free, enlightened and responsible decisions. The view clashes with the doctrine that all events, including our decisions, are predetermined, and therefore that we are not responsible for what we do. I will compare arguments supporting these two positions, and try to show that we are neither free nor responsible, but we are rational and punishable, which is the second best thing.

Körner characterizes several positions in his published lecture, such as material predetermination, logical predetermination and moral indeterminism. "The doctrine of material predetermination is the thesis that the conjunction of all true propositions describing the universe [...] during a particular period of time together with some true timeless principle logically implies all true propositions describing the universe at any subsequent period of time." (Körner 26) The true timeless principle that Körner mentions here is, in his own words, either "a general principle of causality" or "a Leibnizian principle of sufficient reason." (Körner 26).

Logical predetermination presupposes that there are no "propositions referring to a region of the universe during a future period of time [...] which as yet are neither true nor false, but might become either." (Körner 27) In other words, logical predetermination takes all propositions that are true at some to be true at all times.

Moral indeterminism admits that some "as yet undetermined possibilities" can be made into facts by "a person's moral evaluations, decisions and character for which he is, at least partly, responsible." (Körner 27,28).

Körner says that these three positions are not of equal logical strength. "The thesis of material predetermination is stronger than the thesis of logical predetermination. That is to say the former implies, but is not implied, by the latter. Even in the absence of a principle of material predetermination, an omniscient being is conceivable which might know everything that is true, including the description of all past, present and future stages of the universe." (Körner 27) Körner also notes that moral indeterminism is stronger than material indeterminism. (Körner 27).

It will be convenient, before we proceed any further, to have a clear picture of all basic positions and relations involved here. I will assume that for every version of predetermination mentioned by Körner there is a corresponding version of indeterminism, and the views in every pair are strictly complementary and contradictory rather than opposite or mutually exclusive in any other way. The principle of the excluded middle holds in each of the three cases. Predetermination is always a view that events of some sort are fully and unconditionally determined from times immemorial. The indeterminist version of the same position holds only that predetermination is not true, and that at least one event or one type of event (and not necessarily a large region of the universe) are exempted from the universal predetermination. Thus by denying predetermination we obtain an indeterminist position, and by denying indeterminism we obtain a position of predetermination. Consequently, for any two theses, if T1 is stronger than T2 then non-T2 is stronger than non-T1. This usage does not reflect Körner's understanding of the terms, for he defines material indeterminism not as the position complementary to material predetermination but one which "restricts the range of all as yet undetermined possibilities by physical laws, such as a law of non-increasing entropy". (Körner 27)

With the assumption of complementarity of the predeterministic and indeterminist positions and Körner's own statement about their logical strength we can draw a complete picture of the pertinent logical relations: on the one hand, material predetermination implies two positions: logical predetermination and moral predetermination; on the other, moral indeterminism implies material indeterminism, and independently of that, logical indeterminism implies material indeterminism. No other position when true implies any other. If it is granted now that this enumeration of the consequences of true positions is complete, we need not track down the consequences of any position on the assumption that it is false, because each position can be directly translated into a corresponding complementary position, which on the same assumption will be true. Thus, for instance, we do not need to follow the consequences of moral predetermination if we come to believe that it is a false position, for, if we think that it is false, moral indeterminism must be true and we can analyze the logical consequences of the latter position.

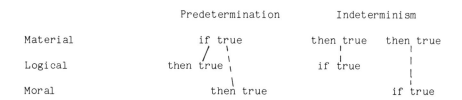

It is clear now that only three positions need to be studied in detail: material predetermination, logical indeterminism and moral indeterminism; others will have been covered without having been addressed. I will begin from logical indeterminism. If this position is found to be sound and compelling we will be permitted to conclude that material indeterminism is true. If not, and I will argue that it is not a compelling position, all problems will be reduced to the controversy between material predetermination and moral indeterminism: if one of the two is true the other cannot be. I will not venture too seriously to solve this controversy, the issue is too complex. But I will argue that the chasm between these two positions is not as deep as often thought, and that a lot of our moral beliefs and attitudes can be salvaged in a world that is made of a tightly packed fabric of causal relations. The logical relation between material predetermination and moral indeterminism, as Körner points out, is that of mutual exclusion, but material predetermination, I will say, is not so devastating for the moral theory and common moral practices as it is often assumed.

Logical Indeterminism

Logical indeterminism is a strong theory, as we have seen, for it can abolish material predetermination independently of any physical, metaphysical or moral considerations. Probably the strongest argument in its favor is an adage variously attributed to Avicenna, Isaac Israeli and

Aristotle, that truth is **adequatio rei et intellectus.** Hence a sentence which describes a future state of affairs in an indeterminist world may be neither true nor false if it describes consequences of an event which is itself undetermined. Let us see what such events are like.

Suppose that uncaused events can happen in nature, and, for instance, some changes in the atmospheric conditions over the Pacific are uncaused. Consequently monsoons cannot be fully predicted. Let us imagine that there will be an important uncaused atmospheric change over the Pacific in 1995 which will double the rainfall in Bombay in 2000. On these assumptions it will be true in 2000 to say "The rainfall in Bombay in 2000 is approximately double of what it has been a decade ago." But can we say that this statement is true already today? A logical indeterminist may argue that will begin to be true at best in 1995, when the increase of the rainfall has been determined by the climate change over the Pacific. Prior to that date there was nothing in the real world which made this statement true. The change of climate was a mere possibility, an event that might or might not happen, being no more than an uncaused, chance occurrence. It seems gratuitous to say that this statement was true by virtue of its correspondence to a certain fact, if that fact was uncertain and belonged to a possible world rather than the real one. Arguably the situation changes in 1995 when the fact in question becomes a prospective consequence of a chance event that has already happened. It belongs to the future real world now, for its causal antecedents are part of the present world.

For an indeterminist the future real world is, so to speak, incomplete, and has to be furnished with the missing elements from the stock of components residing in the possible worlds. Hence if, by definition, a true statement about the real world, or any of its parts, is one that refers to something in that world, then the statement about the rainfall in Bombay will become true only after the climate has changed over the Pacific, and not before. Till then not even an omniscient being can know that the statement "The rainfall in Bombay, etc." is true, because an omniscient being, along the limited mortals, can only know as true what is in fact true, and that statement is not true yet.

The classic refutation of any such attempt to establish temporality of truth comes from Frege, who said that every truth is eternal and independent of whether it is being thought of and of the mental constitution of the one who thinks it. But the indeterminist may respond that Frege did not distinguish between two possible objections to the permanence of truth: the strength of conviction and indeterminacy of the real world. The indeterminist may agree with Frege that the strength of assertion has no bearing on the truth value of the asserted statement: if I tell the truth by mistake, believing that I am telling a lie, what I am saying remains to be true in spite of my evil intentions. If, however, we happen to live in an indeterminate world where some events do not have causes, and for instance the moment when a particular atom emits a quantum of energy is objectively undetermined, then, if the emission of energy initiates a chain of events, the statement predicting this occurrence is neither true nor false prior to that emission, because prior to that emission it is undetermined whether the sequence of ensuing events will happen. Indeterminacy itself is no problem, because we can make precise, descriptive statements about it; we can say, e.g. that the momentum and speed of an electron is undetermined (irrespective of whether it is true or not). But if indeterminate parts of the world can have consequences, or if human freedom is a fact, and whether I will drink a glass of milk tomorrow morning depends entirely on my whim, then prior to the occurrence of the circumstances which constitute the causes of those events sentences predicting their occurrence have no truth value.

This anti-Fregean plea can be simply countered by saying that every doubtful or undetermined fact of the future can still be described by a proposition which truly or falsely predicts its outcome, a statement like "The rainfall in Bombay, etc." The facts to which these statements refer may themselves remain undetermined for some time, but this does not put them out of the scope of the real world and transfer them into the regions of possible worlds. A false statement does not properly describe anything in the real world but is about the real world as much as a true one, or else we would have no grounds for calling it false. Similarly statements about the undetermined future events are about the real world and not about possible worlds even if what they are talking about is still a possibility rather than a tangible fact.

I think that logical predetermination is a sound position and I do not believe that logical indeterminism has any edge over it. I will refrain from saying whether the latter position is feasible or not, because I do not know all its ramifications. I only need to put it out of my way, and I think I have a right to do so, after having shown that the argument from the omniscient being does not cut both ways, that the future world is as real as the present world, that our ignorance of the future truths does not invalidate them as truths, and that the consequences of chance events, if there are such events, are as real as the consequences of the necessary events, if there are such, and therefore statements about either can be true before the events in question come true.

Material Predetermination

The view of material predetermination cannot be easily purged of the strongly metaphysical belief that all events are necessary and describable by general laws. Taking a cue from Kant we might ask, "What are the grounds for saying that every event has a cause?" This proposition is not an empirical generalization, it is hardly a transcendental rule of understanding (unless we swallow Kant's hook, line and sinker), it is not a consequence of statistical reasoning, etc. In short, it is a widespread but self-sustaining prejudice of the scientific mind, quite like the belief in the free will is a widespead but self-sustaining prejudice of the moral mind. There is probably nothing we can do to ultimately disprove one of these prejudices and show the other to be a genuine truth. But we cannot believe in both at the same time. We have to choose between two visions of the world and ourselves, in the first, the world and the mind are strictly determined by prior events, in the second, some fragments of the world and probably most of our minds are indetermined - chance events can happen in the world and free decisions can be reached by the minds. Scientific approach to nature implies acceptance of the first view, our feeling of freedom and control over our bodies suggests that the second view is true. Hence we must either try to reinterpret science in a way which distinguishes between facts and necessary occurrences, caused events and chance, connected parts of the universe and spontaneous happenings unrelated to nothing that happened before them, or we must try to reinterpret man as a natural being, one among many, fully governed by causes and perfectly predictable. Both attempts were made in philosophy many times.

I can say nothing about the correctness of the argument that some objects described by physics, like electrons, are indeterminate and that they are responsible for the occurrence of events without causes. I do not know whether chance events are real or only an illusion of human understanding. The same goes for free will. I even do not know how the ideas that come to my mind are connected by causes, associations or chance. It seems to me that sometimes I control them but on other occasions I feel they come and go as they please. I do not know if a genuine random number generator is possible, and I do not know whether Lande is right when he says that determinism presupposes a demon who must have established with infinite precision the results of every attempt to produce a series of random numbers as well as the results of all conceivable measurements, and thus material predetermination is the most unparsimonious of all known theories (Lande 69f). I will concentrate therefore on one problem only: Assuming that material predetermination is true, how far is it compatible with common moral practices?

Suppose that the human brain works like a computer, that it is equipped with an operational system, an ability to write and execute programs, large memory and a capacity for substituting names or images of objects and events perceived in the world for the variables that are contained in the programs.

On these assumptions we can explain, and developmental psychologists often do, how a child gets its first experience of the world. When a baby sees its mother, her image does not fall into an empty space. The infant already has a vague expectation of a guardian or protector even before it has first seen its mother. Chickens and chimpanzees have that expectation and are ready to follow the first moving, quacking or fuzzy object that they find after birth. Thus in a sense they have an innate variable that stands for their mother or guardian. Such a variable is nothing more than the ability to remember and recognize a particular object and to respond to that object with a behavior that has not been learned. The same holds for sensations. A child does not learn the sensation of pain from its own bruises. The appropriate predisposition of the nerves is already there before the child first hurts itself. This is not to say that all variables are innate. Some must be, but others are learned. I do not believe, for instance, that the idea of a toy is inborn. It is probably formed gradually. A child makes a concept of the first toy as it plays with it, then generalizes this concept to include new toys that have common properties with the first one, and so on. No matter how the concept arises, and whether it is representational or not, it serves as a variable, and these variables are necessary to identify objects in the world around.

I can also imagine that behavioral patterns are remembered as if they were stored in a computer-like memory. There may be programs for sucking, sneezing, blinking the eyes, coughing, crying without a purpose, shouting to bring the mother back, kicking to get rid of wet diapers, wriggling to get out of the warm clothes, etc. Some of these programs may be innate, others may be learned, it does not matter; the important fact is that their execution can be interrupted without getting the child confused. The program of playing with the largest toe can be interrupted by hunger, then the feeling of hunger somehow turns on the program to call the mother. This program is interrupted when the mother comes, and the sucking or munching program is switched on. Why is not the child confused in the process? One answer is that the child does not care which program is on as long as one program is followed by another in a smooth succession, and so that there is no need for a superior, overseeing faculty which controls the shifts from an earlier program to the next. Another answer, which I like better, is that there is in the human brain a special master program that calls on and off the execution of other, task oriented programs. This master program provides for a sense of continuity of one's self and behavior, gives us a stable though not directly recognizable identity that many philosophers tried in vain to describe.

The master program is conceivably responsible for monitoring physical and emotional states of the organism. When a new need or whim arises the master program must first of all decide, if the program being currently executed should be interrupted and a new program turned on, or whether no change is needed. If the new stimulus is strong, or its neglect dangerous - the qualities that the master program must quickly assess - the current program will be interrupted and a new one activated instead. Even without external stimuli the master program must monitor the execution of task programs to see if they are functional and if they produce what they are supposed to produce. Inevitably the master program must be capable of self-correction. If it chooses a wrong task program to obtain a specific end it must be capable of not only stopping it but also of finding or creating a new task program that would better serve the same purpose, and then it must delete a fragment of its own data file which was responsible for the selection of the inefficacious program and write the name of the new program in that place.

Writing a new task program may be a job entrusted to a third kind of program, a learning program. It can be very simple, based on a complete list of body movements and an idea of the aim that needs to be reached. The simplest learning program may work by trial and error. If it is more sophisticated, it may compare the required task with other tasks, and choose a program that is known to be able to complete a task most similar to the one that must be accomplished. If no such program is found, the learning program may divide the task into smaller objectives to be completed by existing programs. If it still is not enough, the program may try to substitute the required task with a related one that can be completed.

If a learning program does not produce required results the task may be abandoned - and that of course happens quite often among humans and other animals - or a fourth kind of program, a representation program, can be turned on. A representation program must be capable of "depicting", "imagining", "mapping out", or "simulating" the situation that needs to be responded to. In the simplest case "simulation" may be no different from imitation. If a person does not know what to do, he or she usually imitates someone who does. If no model can be found, the person may ask questions, read instructions and books, finally, think on one's own. At some point the human computer begins to assemble a comprehensive picture of the world. If no urgent needs interrupt this process, its execution can go on for a long time, and is called studying. Presumably the representation programs paint images or create scenes representing the world out of the information stored in the memory. Elements used in these representations can be connected by relations in which they are found in the real world, and are made to move and influence one another as they do in the world. This inner theatre of the world can include the very person who puts it on stage. He can contemplate and compare the different ways in which he can act, follow the consequences, and, knowing the laws of nature as an average person does, he may discover what means are apt to bring about the ends he would like to produce. If these means require new skills, the learning programs will be able to produce new task programs, possibly very complicated ones, to let the person acquire such abilities as talking to people, driving cars, writing letters, using syringes, making airplanes, fighting in wars, bringing up children, making friends, etc.

The human computer as it has been described here can work in a purely deterministic way. It need not act at random or possess a faculty of free will. The only important condition on which its rationality and, presumably, moral integrity depends, is that it can "stop and think", i.e. frequently monitor the task programs being executed, interrupt them whenever something goes wrong and select new, more functional programs instead. If our computer is a self-correcting device with an ability to choose from the old programs and writing new ones, it will behave like a rational being. The master program will set a goal, largely under the influence of physiological stimuli, it will choose a suitable task program and run it. If the program does not meet the goal, a learning program will be activated, and if this is not enough a representation program will be used until the master program is satisfied or runs out of its options. The computer never has to evaluate anything, choose its own criteria, compare alternatives independently of the criteria set for the purpose to be achieved. To be rational it need not be logical in the sense of being able to perform operations in a sequence determined by patterns of inference rather than by causal relations.

For a computer, learning means having a list of errors and being able to change those parts of currently executed programs which lead to errors. Choosing is being able to review different options and finding which stands foremost in terms of the applicable criteria. Consequently, moral choice involves making a review of all possible courses of action open in the circumstances, having each course of action evaluated by applicable criteria, and selecting one with the highest score. If for some reason this choice is found morally unsatisfactory and another course of action is identified as morally more appropriate, the same task program will be corrected by adding more weight to the considerations that have been neglected or underestimated or by downgrading those considerations which led to wrong choices.

We may call anyone acting along these lines a moral automaton, using a "moral choice program." It works like most known computers, only unlike them, it differentiates between "good" and "bad" states of the world, and between "favored" and "unfavored" states of its own. A "good state of the world" is one which is selected by the "moral choice program" and does not lead to a revision of that program. A "favored state of the computer" is the state that the automaton is preprogrammed to choose unless such choice is overridden by later programs accepted by the master program. The automaton eschews "unfavored states of its own" for its own sake, or, say, because it has the information that choosing them leads to punishment. It may avoid or prevent the occurrence of bad states of the world if the master program assigns some importance to the ranking of different courses of action generated by the "moral choice program." Electronic computers do not have any "favored states of their own". Human beings have them, and these are the states that foster man's well being - plenty of food, good health, a nice job, self-satisfaction, social standing, attractive property, good appearance, etc. Because we have feelings and prefer some situations to others, we can be taught, instructed and even forced to act in different ways. Social censure, legal punishment and moral blame work as very powerful error signals in our personal computers and, in most of us, cause a prompt rearrangement of the ranking of choice criteria. Censure affects us because we want to minimize the 'unfavored states of our own', and we are ready to rewrite our 'moral choice program' for that purpose. Censure also affects us because our programs can be modified and rewritten, because the master program does not allow one program to run forever but makes periodic checks of the entire repertory of tasks programs, has new ones created, and rearranges the strength of the criteria which have led to errors.

There is no evidence, except circumstantial, that human beings are in fact moral automata, but this is largely beside the point. I am not trying to show that human beings are in fact moral automata, it is enough for my purpose to show that they may be, or, in case the human beings have a different constitution, that moral automata are possible. Moral automata have two important properties (1) they are furnished with at least two independent systems of values (unless we agree that perfect egoists having only one system of values are moral automata as well), and (2) they have a master program that reviews and corrects its options and works at the same time when other programs are being executed. The first condition is not difficult to satisfy. In the crudest version of a moral automaton values can be supplied on a list and stored in memory. In more refined versions, values can be produced by organic needs and their cultivation into higher norms through socialization. The second condition is rather puzzling, however. Why are we equipped, if in fact we work like computers, with a master program which controls other programs and continuously evaluates their performance? Why are we not simple, one-track-at-a-time computers, which never look back but blindly go on and on until they are spent? Is there something miraculous about being more complicated? I do not know, but there is definitely something special and most fortunate in this state of affairs. If we did not have that property of self-reflection and self-control we would have looked at ourselves as we look at people with whom we have absolutely no ways of communicating. We could only watch ourselves without being able to influence our behavior in the slightest degree. As things are, we have been nicely endowed by nature with the ability of assessing and guiding our conduct, and it should not matter to us that these faculties have been granted to the computers within us rather than the spiritual entities like the ego, self, mind, a faculty of self-identification, etc.

Moral Indeterminism

The principal tenet of moral indeterminism is, "ought implies can". It stipulates that it is our obligation to do only such morally commendable acts which we can possibly do. Thus, for instance, the moral duty of a doctor to save the patient's life must be construed as a duty to save his life when it is medically possible to do so; and a doctor will not be held morally responsible for the death of a patient who was in a fatal and comatose condition. Although the rule "ought implies can" is fundamental for moral practice, it is not, strictly speaking, one which tells us what is right or what is wrong. It tells us only how to ascribe blame. A doctor who stands helplessly by the side of a dying patient may be doing all that he can do. It would still be better if he could do more; and what he cannot do, namely, effectively help the patient, is still the morally right thing to do if feasible - the patient's death is something bad even if no one can prevent it. But once we have seen that it is inevitable, we must also see that there is no point in putting blame on anyone for his death. In this way, the rule "ought implies can" entails "no one is to blame for what is inevitable," or "one is to blame for what one could not prevent," and this is its main practical sense.

If material predetermination is right, and I think it is prudent to make this assumption (if only to see its consequences), then everything is inevitable, and we are never to be blamed because we could never have acted differently from how we did. But we must be careful to distinguish between questions of blame, responsibility, causal efficacy and inevitability.

If human beings act in some respect like computers, and if moral evaluation is the result of running a "moral choice program", our decisions cannot be said to be free. No amount of probing, what they are not free from, will change the fundamental fact that all decisions are materially predetermined. For those who say "I am free when I follow my feelings", and for those who say "I am free when I can act rationally," for those who say "I am free when I do what is historically necessary" the same answer is right: their decisions are not made here and now, and not by them; their decisions are pre-established from times immemorial, so no matter how they define "being free" they are totally bound and deprived of the faculty of choosing.

A moral indeterminist rejects this belief for purely moral reasons. He is not ready to accept a view of man as a moral automaton. Our rationality and sovereignty, he believes, require freedom of decision; our dignity and moral responsibility require freedom of the will. Without freedom we cannot choose, and consequently we cannot be held responsible, blamed, punished, corrected, persuaded and changed for the better. Without freedom a thief steals because he must, a policeman catches him because he must, a judge judges him because he must and the prison warden keeps him locked up because he must. No one is guided by the right or the wrong, each acts like a puppet. A thief may have a feeling he is doing a bad thing, the judge may have a feeling he is doing something right; but these feelings bear no relation to the causes operating in the situation. The thief steals and suffers, the judge punishes and feels he has done his duty - these feelings, however, like all other events, are evoked by prior circumstances, are predetermined. Given the objectives of the thief and his resistance to moral scruples he must steal, given the training of the judge and his understanding of social order he must punish. None can be right while the other is wrong, says an indeterminist about the determinist position, for neither is capable of studying the situation on its own merits without being influenced in his judgment by his individual causes.

If the last claim were true the indeterminist would indeed have shown that material predetermination turns moral theory into sham. Not only moral considerations would have been caused rather than evoked by an independent moral faculty but even the assessment of the validity of these considerations would be determined rather than justified, i.e. would depend on causes rather than reasons. But this interpretation of predetermination may not be the one that a determinist has in mind. A material predeterminist may believe that moral considerations arise in the mind according to some hidden causal patterns - so far the indeterminist is right. But an assessment of these considerations need not be governed by morally irrelevant causes. Moral evaluation may begin as a causally produced state of mind. Through conditioning can we be taught to regard wife beating, bear baiting, child molesting, etc., as morally wrong. Whenever we meet cases of these kinds we respond with emotional, conscious indignation. A moment later, however, we may reflect on our reaction, and we may put to ourselves a critical question, "Is it really something morally wrong?" At this point two things may happen: either the question itself with contemporary circumstances produces the next response by simple causal connection, or a fairly complicated program of evaluation is activated, and the matter in question is given serious consideration. If we believe that the human mind functions like a computer, it is natural to believe that the second alternative is true. We are capable of impartial assessment of our own views and of our position vis-a-vis other people. Thus the thief and the judge do not know the same thing, and are not equally operated upon by immediate causes. The judge has some conception of social order in which he does not play a central role, while the thief only knows that he needs something badly, but does not propose to make this desire a general guiding principle of action in society at large.

It may be true that even those ideas do not arise freely in the mind according to some logical patterns but are effects of prior causes. Possibly the entire system of social order that the judge is determined to defend unfolds in his mind like a film, frame by frame in causal succession, and when he asks himself the question, "What shall be counted as extenuating circumstances?," he is caused by his mental make up to believe that hunger, state of desperation or failing health reduce the culprit's guilt while drunkenness or leftist convictions do not. If this succession of ideas is so arranged that the causal mechanism always picks up the solution which is also selected by theoretical consideration, the causal mechanism evaluates exactly in the same way as a most conscientious, unbiased and free moral agent would. In principle there is no reason to believe that this theoretical and causal parallelism does not exist. If a computer with a "moral choice program" selects moral answers consistently, if it takes care of all morally relevant circumstances - and the master program would be responsible to see that it does - the theoretical and causal parallelism is ensured as long as the "moral choice program" is allowed to run to the end (i.e. until it finds a solution), and its operation is not interrupted by external causes.

It is not true therefore that if we accept the conditioning theory of learning, interpret man's mind as a computer and identify a moral decision with the final statement of a "moral choice program", then we have to accept moral relativism and the belief that everyone has his own criterion of truth and the right. If it can be shown theoretically that the judge knows more than the thief, is more consistent, farsighted and impartial, then, consequently, we must believe that his moral views are more probably right that the views of the thief, who does not care how the society is organised as long as there are people that he can steal from. Then, if there is a causal mechanism who always makes the same decisions as one who contemplates theoretical issues, the two are indistinguishable for practical purposes. Thus theoretical and causal parallelism, if true, explain how it is possible that certain individuals can think and act as if unconstrained by causes when they reflect on moral questions. The theoretical and causal parallelism makes room for moral theory in a world that is completely predetermined.

This is how we can explain the possibility of moral rationality. We can act rationally due to the theoretical and causal parallelism. Of course we do not know whether and why it exists, if it does. We can imagine, however, how we would feel and act if predetermination without this parallelism were true. First, we would have a short attention span. We would respond quickly and thoughtlessly to all stimuli, and we would never think of the consequences. Second, we would have no "character" in the sense of tending to respond in the same manner irrespective of whether we are rewarded or punished. We would always choose in such a way as to avoid punishment or obtain gratification. Third, we would not profit from talking with other people about our problems or from reading books. We would have no general, all-comprising conception of the world, something that, as I said, can be provided by a "representation program". We would never try to see if our actions are conformable to our beliefs, if they can be criticized or questioned by theoretical arguments, and whether they fit in our picture of the world. But we do all these things, so I can conclude that the simple version of predetermination, without the theoretical and causal parallelism is not true.

There seems to be another sort of parallelism, between mental states and the states of the brain. Basically it can be explained after Holbach as the full dependence of mental states on the modifications of the brain: "the will [...] is a modification of the brain, by which it is disposed to action, or prepared to give play to the organs." (Holbach 11) According to Holbach free will is an illusion, it is a feeling that one can decide and do certain things with one's own body, when in fact the body acts independently of the mind and produces the illusion of being controlled. If I make a simple test and decide to lift my finger, my finger goes up. Some may say, it is a proof that my will controls my body. The result should be interpreted differently, argues Holbach. The body, or its part, the brain, decides to do two things, first, to impress upon my mind that I want and decide to lift my finger, and second, to lift the finger. Hence, in general, there is no direct connection between my decisions and my movements.

One may distrust this interpretation, but I do not think that one can refute it. Rather, as we make more refined tests, we will end up having a more refined description of the preponderance of the brain over the mind. If, for instance, I decide to look at my seconds hand and mark the number of seconds to the full minute in order to raise my finger the same number of seconds after the full minute, the idea is usually carried as successfully as a simple raising of the finger. But the interpretation must be a bit more complex. I must assume that my brain has invented this test, it has decided to check the seconds hand and then wait for the double number of seconds between the seconds hand and the full minute. After that time the brain lifts my finger. All these procedures are coupled with a serial projection of impressions on my mind that *it* is acting rather than my brain. A bit more complicated test would send me to a kindergarten where some children will, as usual, start a fight. I will put them apart and delude myself that I do it out of the feeling of humanity and sense of justice. Instead, it would be my brain's doings which abhors fights among children as much as I do, and graciously lets me believe that my mind fully represents my moral integrity. It is my brain which puts my body in motion and uses my arms to put the children apart, even though I tend to give credit for that to the mental powers. It will be the action of my brain again when I stumble upon the thought that I should feel proud for having done my duty. The brain patronizes me all my life, lurks behind every feeling and decision that I make, shares my feelings and apprehensions, if not directly, then at least in the sense that it takes full account of them and tries to spare me bad experiences whenever doing so is compatible with the feeling of self-esteem it has nurtured in me.

If it is an illusion that our mental self controls our body, it does not have to be an illusion that we possess all the mental faculties and powers that we normally ascribe to ourselves. But they belong to the brain. The self is a result of a fallacy of misplaced subjectivity, I would say. Instead of identifying with the brain and its states, we identify with an intangible stream of impressions and ideas, presumably because we cannot locate our thoughts spatially within our heads. Whatever the reason, we tend to disown our brains and dread to think about ourselves as computers installed in animal bodies.

With the two kinds of parallelism to explain our behavior we can address a number of specific issues concerning the nature of our moral faculties. (1) We are not free, but, as I argued, we do not need freedom to make theoretically correct moral choices. (2) The theoretical and causal parallelism explains how our choices may be right even if they are caused rather than elicited by reasons. (3) We are not responsible for our decisions, because our brain make them as they have been programmed to. The brains are the hardware which has not been chosen or created by human beings. (4) We are corrigible. We respond to persuasion, teaching, criticism and other people's feelings. We were programmed to do so. Our brains can be compassionate, considerate and kind, so if in some of us they are not, it is all right to influence them and try to make them more gentle. Since we are corrigible we are also punishable for doing things that our body has done under the orders from the brain. The punishment should not be directed against our souls or bodies but against our brains, but as the brains cannot be assessed directly as "write protected," some round about way of making them suffer is inevitable. (5) Whatever is found to be right in moral philosophy can be put down as a description of moral order, or a "moral choice program." To some brains such a program may be convincing by its very formulation if it promotes happiness, impartiality, justice, tolerance, compassion. etc. To other brains such a program may seem superfluous. The controversy can only be solved by a tug-of-war. The gentle brains may try to coax the hard brains to become more like them. The hard brains may resist, arguing that man's first duty is to himself, or that history shows the right way, or that every man has his own standards, etc. Possibly some brains are so constructed that they cannot part with egoism or whatever other partisan system of ethics they have created. If so, they must either be tolerated or somehow fended against. (6) The brains have all moral characteristics which we ascribe to ourselves, if the mind and brain parallelism is corect. (7) The decisions made by the brains, although ultimately evoked by external causes, lead to acts which are inevitable in the normal course of events. According to material predetermination my decision to tell the truth which puts me in a bad light today was predetermined aeons ago. But the consequences of my admission would not have obtained had I not told the truth today. Thus our acts make individual but necessary links in the chain of events.

This profile of moral faculties is certainly far from an ambitious one, but I think that it is sufficient for moral integrity and accurate. We are neither free nor responsible, but our acts are causally necessary, and if they are good, they are also instrumental in making good things happen. Thus if moral indeterminism is a view that can be reduced to the claim that "ought implies can" it is not a necessary prerequisite of everyday moral practices. We can act rationally, compassionately and consistently even though we are not free and responsible for our acts. Punishment is probably not the best method to influence a live brain, but if nothing else works, it seems to be the only method left to be tried. And if in the final result some people suffer for what they could not help doing, while others are praised for what they did not choose to do, there is nothing to be done about it short of overhauling the brains of both - a thing we cannot do and should not dare to try if we could.

* * * * * * * * * *

REFERENCES

Baron Holbach, "System of Nature", quoted from Paul Edwards & Arthur Pap, **A Modern Introduction to Philosophy** (The Free Press, New York, 1966).

Stephan Körner, **Abstraction in Science and Morals,** The Twenty-Fourth Arthur Stanley Eddington Memorial Lecture delivered at Cambridge University, 2 February 1971 (Cambridge Universty Press, 1977).

Alfred Lande, "The Case of Indeterminism", in Sidney Hook, **Determinism and Freedom** (New York University Press, 1958).

* * * * * * * * * *

REPLY TO DR. HOLOWKA

Stephan Körner

The main task which Dr.Holowka sets himself in his clear and admirably unpretentious essay is to compare what he calls two "self-sustaining prejudices" and what I should call two theses of "transcendent" or "speculative" philosophy, namely a deterministic view of the world and one which allows for moral freedom. Holowka holds that the two theses are incompatible, but that, if the deterministic thesis is given a certain form then "the chasm between these two positions is not as deep as is often thought" and "material predetermination is not so devastating for moral theory and common moral practices as it is often assumed". In the following remarks I shall make some comments on Holowka's discussion of logical indeterminism and on his conception of man as a moral automaton. I shall then argue that a conception of man as morally free is no less a thought-possibility than Holowka's conception of man as a suitably programmed computer or Leibniz's conception of man as a monad whose actions are predictable and predetermined by its divine programmer. I shall conclude by recalling a pragmatic argument in favour of the transcendent or speculative thesis of human freedom. Since I have no quarrel with Holowka's interpretation of my Eddington Memorial Lecture to which he refers in his paper, I shall feel free to make my points without reference to it.[1] Although Dr.Holowka defends a determinist and Professor Beck an indeterminist conception of freedom, their papers - and hence my replies to them - cover to some extent the same ground.

1. Holowka accepts the doctrine of logical determinism. That is to say he denies that there are open futures or propositions referring to a region of the universe during a future period of time, which are as yet neither true nor false but might become either. The denial is based on Frege's and, it might be argued, Leibniz's doctrine that every truth is eternal - independently of its having or not having a temporal reference. It certainly makes

sense to say that a proposition of the form 'At time t_o an event of type F_o takes place' is true or false at any time. But I also think that it makes sense to accept a logic admitting open futures. An argument to this effect would first of all have to establish the possibility of alternative logics e.g. by characterizing the minimal core of any logic and by showing the feasibility of various peripheral additions including the assumption of open futures. It would then have to adduce reasons for preferring a logic admitting open futures in preference to any logic rejecting this possibility. Certain forms of moral indeterminism would, I think, count as a reason for such a preference.

While I should be prepared to argue that a logic admitting open futures is no less a thought-possibility, than a logic excluding them, it seems in the present context sufficient to separate the issue of a foreseeable or predictable future from the issue of its being predetermined. The nature of the separation can be indicated by recalling the dialogue between Antonius Glarea and Laurentius Valla at the end of Leibniz's **Theodicée**. One of its points is to distinguish between predicting or knowing the future and making it happen. Apollo can consistently say that he knows the future, but that he does not bring it about.[2] Just as there is no inconsistency in assuming that Apollo only knows Sextus' future, but that Jupiter brings it about, so there is no inconsistency in assuming that Sextus' future is in principle predictable and yet to some extent free, i.e. not predetermined. Having thus distinguished between predictability and predetermination, I turn to Holowka's moral automata.

2. I have no objection to Holowka's conception of human beings as moral automata **provided** that it is, as is done by him, put forward as a thought-possibility. A moral automaton "works like most known computers" except that it "differentiates between 'good' and 'bad' states of the world and betweeen 'favoured' and 'unfavoured' states of its own". A 'good state' of the world "is one which is selected by the moral choice program and does not lead to a revision of that program." A "favoured state of the computer" is the "state that the automaton is preprogrammed to choose unless such choice is overridden by later programs accepted by the master program." Holowka elaborates his conception of a moral computer in greater detail. I should have no objection to still further elaborations, e.g. such as are

inspired by Leibniz's conception of man as a monad, programmed by God to function in the best of all possible worlds.

In Leibniz's **Theodicée** the priest Theodorus, who accepts Apollo's distinction between foreseeing the future and bringing it about, nevertheless expresses the plausible view that not Sextus, but Jupiter who programmed him, should be blamed for Sextus's misdeeds. The well-known excuse produced by Athena is, of course, that without programming Sextus as he did, Jupiter could not have created the best of all possible worlds. For those who are dissatisfied with Athena's answer, Leibniz's thesis that Sextus is nevertheless **somehow** responsible for his actions, will be no more satisfying than Holowka's thesis that Sextus is not responsible for them. They might, therefore, be confirmed in their version of indeterminism or continue to look for a version which would be acceptable to them. As far as I am concerned, such a version of indeterminism has to be internally consistent, consistent with the best available empirical knowledge and intellectually and emotionally satisfying. (I do, of course, admit that what is intellectually and emotionally satisfying to me may not be so for others.)

3. Since in my reply to Professor Beck I have indicated my reasons for regarding the assumption of moral freedom as intelligible, as internally consistent and as consistent with Newtonian and post-Newtonian science, I need not repeat them here. But it seems to me appropriate or, at least, permissible to repeat briefly my version of a pragmatic argument in favour of the assumption.[3] The premises are: (1) I and the addressee of the argument agree in our desire that men should conform to a greater, rather than to a lesser degree to their morality; (2) there is no empirical test to establish whether we are or are not morally free; (3) if we are not morally free, then assuming moral freedom makes no difference to the morality or immorality of our actions; (4) if we are morally free, then assuming moral freedom rather than its absence is likely to counteract the tendency to excuse immorality by denying the absence of effective choice and thus to increase the likelihood of greater conformity to one's moral convictions. The pragmatically plausible conclusion is to accept the assumption of moral freedom. I thus agree with Holowka's view that while material determinism and moral indeterminism are incompatible, the usual gap between them can be reduced if one accepts his view of man as a

biological automaton. But I shall continue to prefer the speculative thesis of moral indeterminism until he or somebody else produces a pragmatic argument in favour of material determinism, which for me will outweigh the preceding pragmatic argument against it.

* * * * * * * * * *

FOOTNOTES

1. Cambridge University Press, 1977.

2. "Je sais l'avenir, mais je ne le fais pas", Theodicée 409.

3. See **Metaphysics: Its Structure and Function** (Cambridge, 1984) Ch.17 Section 4.

* * * * * * * * * *

8. LOGIC AND INEXACTNESS

John P. Cleave

A distinctive feature of Körner's work is the novelty and depth of the application of mathematical logic to philosophical problems. His main achievement in philosophial logic is the theory and application - particularly to empirical continua - of a three-valued logic of inexactness. The purpose of this essay is to describe the origin and mathematical development of Körner's logic and its applications. A unifying concept is the notion of **quasi-Boolean algebra,** which is related to the three-valued logic and the calculus of inexact classes much as Boolean algebra is related to the classical, two-valued logic and the traditional calculus of classes.

Körner's main contributions to mathematical logic are rooted in his critique of the positivist view of scientific theories ([K4]), particularly of the role of logic in making theories into instruments of thought and action. A typical defense of the positivist conception has been expressed by Nagel ([24]), p.90). In his examination of the character and cognitive status of theories he outlined a tripartite structure of scientific theories as follows:

> (1) an abstract calculus that is the logical skeleton of the explanatory system and that 'implicitly defines' the basic notions of the system,
> (2) a set of rules that in effect assign an empirical content to the abstract calculus by relating it to the concrete materials of observation and experiment; and
> (3) an interpretation or model for the abstract calculus which supplies flesh for the skeletal structure in terms of more or less familiar conceptual or visualisable materials.

There are two targets of Körner's criticism. The first is the point where thought meets experience - item (2) of Nagel's definition. This is the origin of Körner's three-valued logic and its application to the calculus of inexact classes and the theory of empirical continua. The second is the theoretical edifice - points (1) and (3) - and concerns various constraints on practical and theoretical reasoning. Of these, Körner's discussions of

relevance and of **material necessity** (K6, K8) must be considered as particularly significant. The latter has been developed in [13] and will not be pursued further here. The former, however, is ultimately related to quasi-Boolean algebras and so will be covered in a later section of this paper.

1. Thought and Experience

1.1 Observation language. Philosophers of science (e.g. Campbell [6], Reichenbach [25]) conceive the relation between the abstract calculus of a scientific theory and its empirical content being established by a dictionary of **correlative definitions** [7]. This dictionary translates the **empirical** terms, predicates and propositions of the observation language into theoretical ones which are then organised deductively. It is assumed, in this tradition, that the propositions of the observation language are truth-definite - each is either true or false - and that the abstract calculus is a transmitter of truth. True observation sentences are translated via the correlative definitions into formulas of the calculus. From these the deductive apparatus derives further formulas which, themselves, are correlates of observation sentences (cf.K4,K9). Körner [K4] argues that the observation language considered here is already a **theoretical** construct. It is reached by an **idealisation** of experience.

1.2 Empirical discourse. Empirical discourse includes such notions as empirical individual, empirical class or relation, and empirical continua etc. They are applicable to the familiar, everyday practical world - they enable us to differentiate and integrate experience. Their main characteristics are:
(i) some individuals are indefinite in the sense of not being sharply separable from their background or from other individuals,
(ii) some classes and relations are inexact in the sense of admitting borderline cases, and
(iii) the notion of empirical continuum is relative in the sense that what is continuous under some definite conditions is a discrete collection under others.

> I am inclined to the conjecture that all schemata of empirical differentiation, past or future, exhibit indefiniteness of individuals, inexactness of classes and relations, and relativity of continua. ([K4], p.17)

LOGIC AND INEXACTNESS

The transition from empirical discourse to the observation language entails the replacement of empirical notions by the related **abstract** notions of **definite** individuals, **exact** classes and **absolute** continua. For example, Carnap ([7] p.149) supposes that the observation language for physics is based upon properties of finite space-time domains:

> 'In such and such a place is a horse' means 'such and such a space-time domain has such and such a property'. (p.150)

It is assumed here that properties of space-time domains are exact. The transition from empirical concepts (e.g. 'horse') to the exact notion of space-time domain is amongst the subjects of Körner's analysis.

1.3 A third truth-value. Classical logic deals only with propositions with well-defined truth-values **true** or **false**, and concerns only **exact** properties and relations i.e. those which do not admit borderline cases. It is the logic appropriate to the level of the correlative definitions. Körner claims that there is another logic, capable of representing the inexact, empirical discourse. It can be constructed by adding a third truth-value – **neutral** – to the classical **true** and **false**.

Sometimes the criteria for deciding truth or falsehood of a given empirical sentence S fails to provide an answer. S cannot then be said to have a truth-value. An abbreviated way of acknowledging this situation is to say that S is **neutral**. If S is neutral, its translation into a true (or false) sentence of the observation language for processing via the correlative definitions and the deductive machinery of the scientific theory is problematic. The theory is only applicable to idealised states of affairs described by sentences which are either true or false. To pass from the raw material of experience to the clear, sharp propositions of science is a process of idealisation which can be called the **classical imperative**. It involves responding to the demand to make a **decision** by sharpening the truth-criteria so that neutral empirical propositions become true or false.

The first step in constructing Körner's logic of inexactness is to treat inexactness **as if** it were a truth-value. Other sources of a third truth-value are quantum mechanics [26], partial function logic in the theory of computation [19], theory of descriptions [5] and recursive function theory [17].

1.4 Logic of inexactness: truth-tables. The logic we define is an extension of classical logic. It is itself classical in the sense that it is concerned entirely with truth - definite propositions - those which are known to have one of the truth-values **true, false, neutral**.

Let '0', '1' denote the truth-values **true, false** respectively. Also let 'n' denote neutrality which we now treat **as if** it were a third truth-value. What connectives should be built into the logic? It is quite natural to extend the classical negation, conjunction and disjunction ¬, &, **v**. But there is more than one way of making such an extension. Körner uses two principles here. Firstly, p & q is false if one of p,q is false, and p **v** q is true if one of p,q is true. Secondly, once a compound proposition has been evaluated as true or as false, this decision in unchanged when any neutral component is sharpened to a classical value (the principle of **permanence** [K4]). The truth-tables are therefore as shown in Fig.1. The logic based on these truth-tables was first defined by Moisil in 1935 ([20]).

p	¬p
0	1
n	n
1	0

&	0	n	1
0	0	0	0
n	0	n	n
1	0	n	1

v	0	n	1
0	0	n	1
n	n	n	1
1	1	1	1

Fig.1. Strong truth-tables.

These tables are exactly the "strong truth-tables" ([17] p.334) used by Kleene in the theory of partial recursive functions. A topological interpretation of them was given in [9] wherein the permanence principle is related to continuity. The logic based on them has been used by Hájek [15] in the problem of automating the evaluation of empirical data. Körner made it the foundation of his theory of inexactness and empirical continua.

For later use, we add a second neutral, m, to the strong truth-tables to give the tables of Fig.2.

p	¬p
0	1
n	n
m	m
1	1

&	0	n	m	1
0	0	0	0	0
n	0	n	0	n
m	0	0	m	m
1	0	n	m	1

v	0	n	m	1
0	0	n	m	1
n	n	n	1	1
m	m	1	m	1
1	1	1	1	1

Fig. 2.

The logic based on these tables was first defined by Muškardin [22] in connection with his theory of entailment which will be considered later.

1.5 Logic from truth-tables. The traditional development of a logic from a set of truth-tables proceeds via the stages:
(1) from a formal language construct the set W of formulas,
(2) from the truth-tables define how to compute the truth-values of compound formulas from the truth-values of atomic formulas,
(3) construct the relation of logical consequence, C, via the Bolzano-Tarski definition [29]. The structure ⟨W,C⟩ is the **logic**.
(4) from the notion of logical consequence construct the relation ≡ of logical equivalence. The equivalence classes of formulas then form an algebra satisfying all those identities satisfied by the algebra of truth-values.

For stage 1 consider a propositional language based on a set P of propositional variables, connective signs &, **v**, ¬ and two propositional constants **0, 1**. The set W of formulas is constructed in the usual way. (It will, however be convenient sometimes to write 'AB' as an abbreviation of 'A & B' and '$\bar{A}$' as an abbreviation of '¬A'). The connectives can play the role of functions. Thus:
$$\mathbf{v}(A,B) = (A \mathbf{v} B), \quad \&(A,B) = (A \& B), \quad ¬(A) = ¬A.$$
Step 1 is completed by constructing the algebra W of formulas:
$$F = \langle W; \mathbf{0,1,\&,v,¬} \rangle$$
The semantics of the language is constructed in step 2. The method of assigning truth-values to compound formulas must be defined. It is useful to treat the truth-values as algebras, the operations of which are given by

the truth-tables. The classical truth-tables, the strong truth-tables (Fig.1) and the truth-tables of Fig.2 give the algebras

$$B_2 = \langle \{0,1\} \quad ; \quad 0,1,\&,\mathbf{v},\neg \rangle$$
$$N_3 = \langle \{0,n,1\} \quad ; \quad 0,1,\&,\mathbf{v},\neg \rangle$$
$$Q_4 = \langle \{0,n,m,1\} \quad ; \quad 0,1,\&,\mathbf{v},\neg \rangle$$

The operations of conjunction and disjunction, and the constants 0,1 make these algebras into lattices ([4]) with maximum element 1, and minimum element 0. B_2 is, of course, a Boolean algebra. As the strong truth-tables extend the classical ones, B_2 is a subsystem of N_3. N_3 is also a subsystem of Q_4.

To achieve some generality in step 2 we therefore consider a finite algebra T = $\langle A; \mathbf{0,1,\&,v,\neg} \rangle$ of truth-values. It will be assumed that under the operations of &, **v**, the algebra T is a lattice with maximum element **1** and minimum 0. As usual ([4], p.18) the lattice order $\leq$ is defined by

$$x \leq y \leftrightarrow x \mathbf{v} y = y$$

We observe that in B_2 $0 \leq 1$, in N_3 $0 \leq n \leq 1$ and in Q_4 $0 \leq n, m \leq 1$ (here n and m are incomparable).

A **valuation in T** is a mapping $\tau: P \cup \{0,1\} \rightarrow A$ such that $\tau(0) = 0$ and $\tau(1) = \mathbf{1}$. "Val(T)" denotes the set of all such mappings. Each valuation $\tau \in$ Val(T) can be extended via the operations &, **v**, $\neg$, of T to a homomorphism, which we again denote by "τ", of W to T. If T = B_2, the homomorphism is computed by the usual classical truth-tables which define the operations &, **v**, $\neg$ on $\{0,1\}$. Similarly, if T = N_3, the evaluation of compound formulas is effected via the truth-tables of Fig.1 which define the operations &, **v**, $\neg$ on $\{0,n,1\}$.

A notion of **logical consequence** can be formulated which is a generalisation of the classical Bolzano-Tarski definition. In classical logic, a formula A is a logical consequence of the set X of formulas (written as "A $\in$ C[B_2](X)" if every valuation in Val(B_2) which makes all formulas in X true, also makes A true. This is generalised to:

$$A \in C[T](X) \leftrightarrow \tau(A) \geq \min(\tau X), \text{ for all } \tau \in \text{Val(T)}. \quad (1)$$

where min(τX) is defined to be 1 if X = ϕ. This definition relies on the lattice ordering of the truth-values. It is easily seen to coincide with the classical concept when T = B_2. The **logic of T** is the structure L(T) = $\langle W; C[T] \rangle$. We remark that the consequence function C[T] satisfies the condition:

LOGIC AND INEXACTNESS

for all $X, Y \subset W$

(i) $X \subset C(X)$

(ii) $C(X) \subset C(X \cup Y)$,   (2)

(iii) $C(C(X)) \subset C(X)$, and

(iv)   if $A \in C(X)$ then $A \in C(Z)$ for some **finite** subset Z of X.

$L(T)$ is therefore a **deductive system** ([10], [29]).

To compare logics it is useful to note that if T_1 is a subsystem of T_2 then the $Val(T_1) \subset Val(T_2)$ and hence from (1)

$$C[T_1](X) \supset C[T_2](X). \qquad (3)$$

For instance, as B_2 is a subsystem of N_3 and N_3 is a subsystem of Q_4

$$C[B_2](X) \supset C[N_3](X) \supset C[Q_4](X). \qquad (4)$$

Note also that by (1)

$$A \in C[T](\phi) \leftrightarrow \tau(A) = 1 \text{ for all } \tau \in Val(T). \qquad (5)$$

The traditional treatment of classical logic, $L[B_2]$, singles out the class of tautologies, i.e. $C[B_2](\phi)$, for special treatment since, in a way, they encapsulate the relation of logical consequence: i.e.

$$\neg A \vee B \in C[B_2](\phi) \leftrightarrow B \in C[B_2](\{A\}). \qquad (6)$$

This is the reason for introducing the material implication connective, $\supset$, into classical logic. $A \supset B$ always has the same truth-value as $\neg A \vee B$. Hence by (6)

$$A \supset B \in C[B_2](\phi) \leftrightarrow B \in C[B_2](\{A\}).$$

Thus the meta-relation of logical consequence is exactly reflected by a connective in the language itself.

Let us call a formula **pure** if neither **0** nor **1** is a subformula. In $L[B_2]$ there are many pure tautologies; in $L[N_3]$ there are none since every pure formula takes the value n when all atomic formulas take this value. Nevertheless, it is possible to define the classical tautologies in terms of the valuations in N_3:

$$C[B_2](\phi) = \{A \in W: \tau(A) > 0 \text{ for all } \tau \in Val(N_3)\}. \qquad (7)$$

But, better still, they can be defined in terms of the three-valued logical consequence:

$$C[B_2](\phi) = C[N_3](\{p \vee \bar{p}; \ p \in P\}). \qquad (8)$$

There are two kinds of axiom system for classical logic. One type formulates Hilbert-type systems in which the derived formulas are precisely the tautologies. The other kind is the Gentzen-type of system in which the derived sequents $A_1, \ldots, A_n \rightarrow B$ are exactly the pairs $(\{A_1, \ldots A_n\}, B)$ such that $B \in C[B_2](\{A_1, \ldots, A_n\})$. In view of the absence of pure tautologies in the three-valued logic there is no point in trying to use Hilbert-type systems for $L[N_3]$. Gentzen-type systems can, however, be used to axiomatise the notion of logical consequence in $L[N_3]$. This has been done in [8] where a completeness proof was given for such an axiomatisation of the three-valued predicate calculus.

1.6 Algebra of logic. The fourth step in the development of a logic is to pass from **logical consequence** to **logical equivalence**. Let T be a lattice of truth-values. Two formulas $A, B \in W$ are **logically equivalent** if they have the same logical consequences:

$$A \equiv B \bmod(T) \leftrightarrow C[T](\{A\}) = C[T](\{B\}).$$

Thus, by (1), A and B are logically equivalent if, whatever truth values are assigned to atomic formulas, A and B always have the same truth-values. Hence:

$$A \equiv B \bmod(T) \leftrightarrow \tau(A) = \tau(B) \text{ for all } \tau \in \mathrm{Val}(T). \qquad (9)$$

Note that $\equiv \bmod(T)$ is a congruence on W so that the **algebra of T, $\mathbf{A}(T)$**, can be defined as the reduced algebra

$$F \bmod(T)$$

It can easily be established that if T_1 is a subsystem of T_2 or a homomorphic image of T_2 then

$$A \equiv B \bmod(T_2) \rightarrow A \leftrightarrow B \bmod(T_1). \qquad (10)$$

Hence:

If T_1 is a sub-algebra or a homomorphic image of T_2 then

$\mathbf{A}(T_1)$ is a homomorphic image of $\mathbf{A}(T_2)$

For example, since B_2 is a subsystem of N_3 and N_3 is a subsystem of Q_4, $\mathbf{A}(B_2)$ is a homomorphic image of $\mathbf{A}(N_3)$ which itself is a homomorphic image of $\mathbf{A}(Q_4)$. Further, the algebra of T inherits many of the properties of the system of truth-values:

T and $\mathbf{A}(T)$ satisfy exactly the same identities.

What are the algebras $\mathbf{A}(T)$? What are the identities which hold in B_2, N_3, Q_4?

1.7. Quasi-Boolean algebras. An algebra $\langle A; 0, 1, \cap, \cup, ' \rangle$ which satisfies the identities K1-K8 below is a **quasi-Boolean algebra** (QBA).

K1. $x \cap x = x = x \cup x$.
K2. $x \cap y = y \cap x$, $x \cup y = y \cup x$.
K3. $x \cap (y \cap z) = (x \cap y) \cap z$, $x \cup (y \cup z) = (x \cup y) \cup z$.
K4. $x \cap (x \cup z) = x = x \cup (x \cap z)$.
K5. $x \cup (y \cap z) = (x \cup y) \cap (x \cup z)$, $x \cap (y \cup z) = (x \cap y) \cup (x \cap z)$.
K6. $(x \cap y)' = x' \cup y'$, $(x \cup y)' = x' \cap y'$.
K7. $x'' = x$.
K8. $0 \cap x = 0$, $1 \cup x = 1$.

Thus, a QBA is a distributive lattice (K1-K5) with maximum and minimum elements (K8) satisfying de Morgan's laws (K6).

B_2, N_3, and Q_4 are QBA's. Since QBA's are lattices they can be represented by Hasse diagrams [4]. The action of the function $(.)'$ is shown by a dotted line joining x to x'. The QBA's with 2,3 or 4 elements are shown in Fig.3.

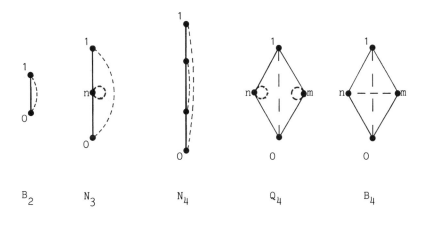

B_2 N_3 N_4 Q_4 B_4

Fig.3. QBA's

B_2, N_3, N_4 and B_4 also satisfy the identity:
K9. $x \cap x' \leq y \cup y'$.

QBA's satisfying K9 are known as normal **quasi-Boolean algebras**. Q_4 is not a normal QBA. QBA's satisfying

K10. $x \cap x' = 0$, $y \cup y' = 1$

are **Boolean algebras**. Thus, every Boolean algebra is a normal Boolean algebra. B_2 and B_4 are Boolean algebras. N_3 and N_4 are normal QBA's but not Boolean algebras.

The term "quasi-Boolean algebra" was introduced by Białynicki-Birula and Rasiowa in [3]. A.Monteiro [21] named these algebras "De Morgan algebras"; he reserved the term "De Morgan lattices" for algebras defined by K1-K7. Several authors have stated that the Morgan lattices were examined by Moisil in [20], but this claim is not firmly based. De Morgan lattices were studied by Kalman [16] under the term "distributive i-lattices": in [10] they were called K-algebras. Distributive i-lattices satisfying K9 were called "normal distributive i-lattices" in [16]. A.Monteiro called them "Kleene lattices" (cf. [8]) and used the term "Kleene algebra" for "normal quasi-Boolean algebra".

It has been proved that:

(1) Every Boolean algebra is a sub-direct power of B_2,
(2) Every normal QBA is a sub-direct power of N_3,
(3) Every QBA is a sub-direct power of Q_4.

(For (1) see [4] and for (2) and (3) see [16].) Hence the identities that hold in every QBA (normal QBA, Boolean algebra, respectively) are exactly those that hold in Q_4 (N_3, B_2, respectively). Further, $A(Q_4)$ is a QBA, $A(N_3)$ is a normal QBA and $A(B_2)$ is a Boolean algebra. The three logics $A(Q_4)$, $A(N_3)$ and $A(B_2)$ differ by their concepts of logical equivalence. Thus, if $p, q \in P$, where $p \neq q$, then $p\bar{p} \equiv q\bar{q} \bmod(B_2)$ but not $\bmod(N_3)$ and $p\bar{p}$ v $q\bar{q} \equiv q\bar{q} \bmod(N_3)$ but not $\bmod(Q_4)$. Nevertheless the algebras are closely related. It follows from (10) that $A(B_2)$ is a homomorphic image of $A(N_3)$ which itself is a homomorphic image of $A(Q_4)$.

QBA's also occur in areas other than logic ([11]). The following construction of QBA's from topological spaces supports their application to finite continua. Let X be a topological space and let 'X' denote the set of ordered pairs (x,y) of closed sets x,y such that $x \cup y = X$ and $x \cap y$ is a closed boundary set ([18]). Define operations $\cup, \cap, '$ on X and constants 0, 1 by:

$$(x,y)' = (y,x)$$
$$(x,y) \cup (s,t) = (x \cup y, s \cap t)$$
$$(x,y) \cap (s,t) = (x \cap y, s \cup t) \qquad (11)$$
$$0 = (\phi, X)$$
$$1 = (X, \phi).$$

'K(X)' denotes the algebra so formed. It follows from well-known results ([4]) p.51, [18] p.125) that K(X) is a normal QBA.

1.8 Inexact classes. Körner's theory of inexact classes can be defined as follows. Let A be a given non-empty domain of (exact) individuals. The classical conception of a **property** of these individuals allows us to say of a given individual **a** that '**a** has P' is true or '**a** has P' is false. But Körner allows that '**a** has P' could be neutral. Thus P

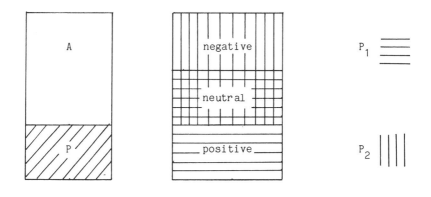

Fig.4.

determines a function. $\psi: A \to \{0,n,1\}$ which in the **characteristic function** of P. The **neutral cases** of P are those members x of A such that $\psi(x) = n$; the **positive** and the **negative** cases are those for which $\psi(x) = 1$ and $\psi(x) = 0$ respectively. So we define an **inexact class** (of A) to be a function $\psi: A \to \{0,n,1\}$. Amongst the inexact classes are the **exact classes**, namely, those inexact classes ϕ such that $\phi(x) \in \{0,1\}$ for all $x \in A$. Körner ([K4] p.42) defines the operations of union, intersection and complementation on inexact classes by the obvious analogy with the classical procedure. Let ϕ, ψ be inexact classes. Then $\phi \cup \psi$, $\phi \cap \psi$ are the inexact classes defined from the strong truth-tables by

$(\phi \cup \psi)(x) = \phi(x) \mathbf{v} \psi(x)$
$(\phi \cap \psi)(x) = \phi(x) \& \psi(x)$
$(\phi')(x) = \neg \phi(x)$

for all $x \in A$. The universal class **1**, and the null class **0** are obviously defined by

$$\mathbf{1}(x) = 1, \mathbf{0}(x) = 0 \text{ for all } x \in A.$$

The Venn diagram is a useful heuristic device for understanding relations between sets. The domain A is represented by a plane region. A subset P of A is then represented by a shaded area included in A and the complement of B (with respect to A) by the unshaded area. Thus an exact class can be considered as a dichotomy of A. An inexact class, P, on the other hand, gives rise to a trichotomy of A - three regions must be drawn representing the positive, negative and neutral cases (Fig.4). It is sufficient to consider the ordered pair (P_1, P_2) of sets of non-negative cases and non-positive cases of P, respectively. Thus, $P_1 \cap P_2$ is the set of neutral cases. In fact, the mapping $\phi \rightarrow (\{x; \phi(x) \neq 0\}, \{x; \phi(x) \neq 1\})$ is a bijection of inexact sets of A and ordered pairs (X,Y) of subsets X,Y of A such that $X \cup Y = A$. If operations of union, intersection and complementation on these ordered sets are defined as in (11), the bijection becomes an isomorphism between the algebra of inexact classes with operations defined by (12) and the algebra of ordered pairs of sets with operations defined by (11).

Körner noted that the inexact classes behave like a Boolean algebra. More exactly, it follows from (11) or (12) that they form a normal QBA. Furthermore, every normal QBA is isomorphic to a field of inexact sets, i.e. a class of inexact sets closed under the operations $\cup, \cap, '$. This structure theorem is analogous to the well-known theorem that every Boolean algebra is isomorphic to a field of (exact) sets ([4] p.159).

1.9 The centre of a QBA. Körner used inexact classes to analyse empirical continua ([4] Ch.4). An important role is played in this work by the idea of the neutral cases of an inexact class P constituting a sort of boundary between the positive and negative cases. An individual suffering **continuous** qualitative change can start by having the property P and end by having the property non-P but in between, at some time, enter a transitional form in which it is neither P nor non-P.

Let P = (U,V) be an inexact class (so that $U \cup V = A$). The neutral cases of P are $U \cap V$, which is a different **type** of entity from P. It can be made trivially into an inexact class by pairing with A to give $(U \cap V, A)$ which

is an inexact class with no positive cases. By (11)
$$P \cap P' = (U \cap V, A).$$

The **boundary** ∂P of P is therefore defined by
$$\partial P = P \cap P'. \qquad (13)$$
Dually, the **coboundary**, P, of P is defined by
$$\delta P = P \cup P'. \qquad (14)$$
It is clear that these definitions can be generalised to arbitrary QBAs.

Definition. Let K be a QBA and $x \in K$.
(i) $\partial x = x \cap x'$, $\delta x = x \cup x'$
       $\partial K = \{\partial x : x \in K\}$, $\delta K = \{\delta x \ ; \ x \in K\}$.
(ii) (∂K) is the ideal generated by the set ∂K of boundary elements of K,
(iii) $Z(K) = K \bmod (\partial K)$ is the **centre** of K.

(For the term 'ideal' and the role of ideals in lattice theory see Birkhoff [4]).

The boundary elements make an important link with Boolean algebras. If we could somehow ignore neutral elements we might return to two-valued logic and exact sets. The precise sense of this is the theorem of Moisil ([20] p.95)

     $Z(K)$ is a Boolean algebra.

In particular, the classical two-valued logic is the centre of the three-valued logic, i.e.

     $Z(\mathbf{A}(N_3))$ is isomorphic to $\mathbf{A}(B_2)$.

(see [9]).

1.10 Empirical continuity. Empirical continuity is a property of finite sets of attributes of attributes or classes and those derived from them by purely logical means, based upon their continuous connection. Körner's method of investigating this notion is founded on the theory of inexact classes, the fundamental assumption being that

> ....an immediate transition from one class to another is discontinuous unless it is a kind of 'merging' of the two or a 'shading into each other' which presupposes that the two classes not only have neutral candidates, but also that some of these are common to the two classes. ([K4] p.50)

The analysis of empirical continuity

....must, therefore, be attempted not in terms of relations between an infinite number of exact classes or definite individuals, but in terms of relations between a finite number of inexact classes or indefinite individuals. ([K4] p.50).

Körner's concept of an empirically continuous series of classes P_1, $P_2,...P_n$ includes the notion of a class being divisible by its neighbours. The concept of continuity therefore relates to the totality of classes constructed by logical means from $P_1, P_2,...,P_n$, i.e. the QBA generated by them. This suggests that the notion of empirical continuity can be extended to arbitrary QBAs. By taking this step we can preserve the spirit of Körner's inquiry and relate it to the topological notion of connectedness.

The notion of connectedness is imported into the theory of QBAs via the QBA K(X) of a topological space X which was introduced in Section 1.7.

Let $(c,d) \in K(X)$. Then c is **connected if**, and only if, for all x, y $\in$ K(X)
$(c,d) = x \cup y$ & $x \cap y = 0 \rightarrow x = 0$ or $y = 0$.

This result justifies the

Definition. Let K be a finite QBA.

(i) An element z of K is **connected** if for all x, y $\in$ K
$x \cup y = z$ & $x \cap y = 0 \rightarrow x = 0$ or $y = 0$.

(ii) K is connected if every element of δK is connected.

The topological notion of connectedness appears as a limiting form of connectedness of finite QBAs in the following way. Let D be an n-cell in R^n. There are many ways that D can be divided into ever smaller 'pieces'. For each n let S(n) be a **finite** collection of n-cells such that

(i) $\bigcup S(n) = D$,

(ii) for all X, Y $\in$ S(n) such that $X \neq Y$, $X \cap Y$ is an m-cell for some m < n or $X \cap Y = \phi$,

(iii) for each X $\in$ S(n), diam(X) < n^{-1}.

'L_n' denotes the lattice generated by unions and intersections of the cells in S(n); 'B_n' denotes the set of elements of L_n of dimension less than n. Now let C be a closed subset of D and $L_n(C) = \{X \cap C;\ X \in L_n\}$. $L_n(C)$ under the operations of union and intersection is a distributive lattice of closed subsets of C. Let $B_n(C) = \{Y \cap C: Y \in B_n\}$. Finally, let $K_n(C)$ be the set of ordered pairs (x,y) where x, y $\in L_n(C)$, $x \cup y = C$ and $x \cap y \in B_n(C)$. Define operations $\cap$, $\cup$,' on $K_n(C)$ by (11). This makes $K_n(C)$ into a normal

QBA and
- (i) If C is topologically connected then for all k, $K_k(C)$ is a connected normal QBA,
- (ii) if C is not connected then there exists k such that $K_m(C)$ is not connected for all $m \geq k$.

In this sense the topological notion of connectedness is an asymptotic form of QBA connectedness.

Finally we can define the sense in which the three-valued logic is connected in contrast to the classical logic.

Let P be a finite set of propositional variables. Then every element of $A(N_3)$ is connected so that $A(N_3)$ is connected.

$A(B_2)$ is not connected: the only connected elements are the zero element and the atoms.

* * * * * * * * * *

2. Theoretical Edifice

Certain intensional concepts frequently occur as ontological and deontological constraints on theoretical and practical reasoning. Körner's work in these areas (e.g. [K7]) is based on the systematic use of the familiar syntax and semantics of classical logic. There is an implied rejection of the method of postulation which supports much of the industry of philosophical logic:

> The method of 'postulating' what we want has many advantages: they are the same advantages of theft over honest toil. Let us leave them to others and proceed with our honest toil. ([27] p.71)

2.1 Relevance. Certain kinds of implications, such as $A \& \neg A \rightarrow B$ and $B \rightarrow A \vee \neg A$ are sometimes regarded as fallacious or paradoxical, even though classically correct, on the grounds that $A \& \neg A$ (or $A \vee \neg A$) is irrelevant to the arbitrary proposition B. Such paradoxes were widely discussed after C.I. Lewis proposed his system of strict implication. Various attempts have been made to characterise a notion of valid implication by constructing new logics based on posited properties of 'entailment'. An admirable, well-documented paper is by Anderson and Belnap [1]. Körner avoids the postulational approach and psychological criteria by seeking the source of paradox in conditions which are **definable within** classical logic:

> ... if somebody tells me that from the axioms of Euclidean
> geometry he cannot only deduce Pythagorus' theorem but also
> the alternative of Pythagorus' statement and the statement
> that Pythagorus' mother was Viennese, and then by contra-
> position, antilogism etc, some other odd deducibility
> statements, I will tell him not to worry because he can
> separate the odd from the other deducibility statement by a
> logical criterion rather than a Gricean psychological one.
> The logical criterion would be my notion of **relevance** or
> something very near it. [K5]

Formulas which have an irrelevant subformula were called **slack** in [12]; non-slack formulas were called **rigid**. In classical logic a tautology $A \supset B$ can be regarded as a **valid entailment** if $A \supset B$ is rigid. Körner's project was to study the closure properties of rigid propositions and rigid tautologies rather than lay down **a priori** postulates for "valid entailments".

The basic principle of relating entailments to rigid propositions can be traced to a notion of 'pure entailment' in [K1]. It occurs more explicitly in [K2] where Körner defines an 'essential' logically necessary proposition as one not containing an 'inessential component'.

Körner has suggested two notions of 'irrelevant'. These are conveniently defined using the following notation of Schütte [28]. 'F(A)' denotes a formula F in which an occurrence of the formula A is distinguished. Given F(A), 'F(¬A)' denotes the formula obtained from F(A) by replacing the distinguished occurrence of A by ¬A. Also, 'F(.)' denotes the sequence of signs arising from F(A) when the following rules are applied:

1) A in F(A) is deleted,

2) If in F(A) a sign is removed which immediately follows ¬, then the negation sign is also removed,

3) If F(A) contains a component (B & C), (C & B), (B v C), (C v B) in which B must be removed, then this component is replaced by C.

    Definition.

    (i)   An occurrence of a subformula X in the formula F is **irrelevant** if $F(X) \equiv F(\overline{X}) \mod(B_2)$. It is **d-irrelevant** if $F(x) \equiv F(.) \mod(B_2)$.

    (ii)   A formula is **slack** (d-slack) if it contains an irrelevant (d-irrelevant, respectively) subformula and **rigid** (d-rigid, respectively) otherwise.

    (iii) R,S, d-R, d-S denote the classes of rigid-,

slack-, d-rigid-, d-slack- formulas respectively.

T denotes the class of tautologies.

Atomic formulas are rigid and d-rigid. (p **v** $\bar{q}$)p is rigid but not d-rigid. p ⊃ p(q **v** $\bar{q}$) and p **v** $\bar{q}$q .⊃ p are rigid tautologies but are not d-rigid. pq **v** pr .⊃. p(q **v** r) and p(q **v** r) .⊃. pq **v** r are both rigid tautologies, but pq **v** pr .⊃. pq **v** r is a slack tautology.

>Definition. A formula F is **minimal** if every formula G
>which is logically equivalent to F is also not shorter
>than F. M denotes the set of minimal formulas.

It can be shown that

$$M \subset d\text{-}R \subset R \text{ and } M \neq d\text{-}R \neq R.$$

Clearly, every formula is equivalent to a minimal formula. Hence, every formula is equivalent to a d-rigid formula.

To some extent d-R and R satisfy the same closure properties with respect to the Boolean algebra laws:

If K = R or d-R then

(1)   A ε K → ¬A ε K
(2)   F ε K and G a subformula of K → G ε K
(3)   F(AB) ε K   ↔   F(BA) ε K
(4)   F(A **v** B) ε K   ↔   F(B **v** A) ε K
(5)   F((AB)C) ε K   ↔   F(A(BC)) ε K
(6)   F((A **v** B)C) ε K   ↔   F(AC **v** BC) ε K
(7)   F(¬¬A) ε K   ↔   F(A) ε K
(8)   F(¬(AB)) ε K   ↔   F(¬A **v** ¬B) ε K
(9)   F(¬(A **v** B)) ε K   ↔   F(¬A . ¬B) ε K
(10) F((A **v** B)(A **v** C)) ε K →   F(A **v** BC) ε K
(11) F(AB **v** AC) ε K   →   F(A(B **v** C)) ε K.

Further,

(12) F(AA) ε R   → F(A) ε R.
(13) F(A **v** A) ε R   → F(A) ε R.

These results were proved in [12].

2.2 Entailment.

Körner's diagnosis of paradox in implications is, partly at least, captured by the notion of rigid tautology.

Definition.

(i) The relations $\vdash$, $\vdash_d$ between formulas is defined by

$A \vdash B \leftrightarrow A \supset B \in R \cap T$

$A \vdash_d B \leftrightarrow A \supset B \in d\text{-}R \cap T$.

If $A \vdash B$ we say that A **entails** B; if $A \vdash_d B$ we say that A d-entails B.

(ii) d-Ent = $\{(A,B); A \vdash_d B\}$

Ent = $\{(A,B); A \vdash B\}$

Imp = $\{(A,B); A \supset B \in T\}$.

Clearly

d-Ent $\subset$ Ent $\subset$ Imp and d-Ent $\neq$ Ent $\neq$ Imp.

Implications which are most obviously paradoxical are those in which either premiss or conclusion are tautologous or contradictory (e.g. $A\bar{A} \to B$ and $B \to A \vee \bar{A}$). Entailments are non-paradoxical in this sense since if $A \vdash B$ then neither A,B, A nor B is a tautology. Hence, if A is a tautology, there does not exist a formula B such that (A,B) $\in$ Ent or (B,A) $\in$ Ent. (This is a stronger criterion than that of Smiley which is referred to in [1] p.10).

The classes of entailments have the following closure properties:

Let E = Ent or d-Ent. Then

(i) If (A,B) $\in$ E and (A,C) $\in$ E then (A,BC) $\in$ E,

(ii) If (A,C) $\in$ E and (B,C) $\in$ E then (A $\vee$ B, C) $\in$ C,

(iii) ((A,B) $\in$ E if, and only if, $(\bar{B},\bar{A})$ $\in$ E.

The converses of (i) and (ii) do not hold. It is quite easy to find 'desirable' properties of logical consequence which fail to hold for the above defined notions of entailment. Thus, (AB,A) $\notin$ Ent and (A,A $\vee$ B) $\notin$ Ent. Further, entailment is not transitive: for example, pq $\vee$ pr .$\supset$. p(q $\vee$ r) and p(q $\vee$ r) .$\supset$. pq $\vee$ pr are both rigid tautologies but pq $\vee$ pr .$\supset$. pq $\vee$ r is a slack tautology. This confirms the opinion of Lewy, Geach and Smiley expressed in [1] p.11.

2.3 Multi-subject semantics. Another source of paradox can be traced to the occurrence within the same sentence of expressions having entirely different meanings. This is clearly acknowledged in the above quotation [18]. It was generalised by Muškardin to provide a semantic foundation for a logic of entailment [22].

Consider k distinct subjects of discourse $U_1,\ldots,U_k$. Let us resolve that the classical truth-values are different in distinct subjects. Thus, if $i \neq j$, then truth (or falsehood) in U_i - denoted by "1_i" (or "0_i", respectively) - is different from truth (or falsehood) in U_j. Within each subject U_i the truth-values $B_2 = \{0_i, 1_i\}$ form a Boolean algebra under the usual operations $\cap, \cup, '$. Now consider the sentence $\neg(A \& (B \vee C))$ in which A is true about U_1, B is false about U_2 and C is true about U_3. The truth-value of the sentence would naturally be $(1_1 \cap (0_2 \cup 1_3))'$. But how should this be computed? The problem is clearly to extend the functions $\cap, \cup, '$ which are defined on **each** B_2, to the domain

$$\bigcup_{r=1}^{k} B_2^{(r)}.$$

It is quite natural to take the compound truth-values in the free quasi-Boolean algebra product, T_k, of the $B_2^{(i)}$ (see [14] for the definition of 'free product'). Thus in T_k

$$(1_1 \cap (0_2 \cup 1_3))' = 1_1' \cup (0_2 \cup 1_3)' = 0_1 \cup (1_2 \cap 0_3).$$

T_k is a finite QBA. The Haase diagram of T_2 is shown in Fig.5. The elements of T_k are the **k-subject truth-values**. The 4-element QBA Q_4 (Fig.2) enters the development now. For if $k \geq 2$ we can put $\rho(1_1) = \rho(0_1) = n$ and $\rho(1_2) = \rho(0_2) = \ldots = \rho(1_k) = \rho(0_k) = m$ so that ρ extends to a homomorphism of T_k onto Q_4. Hence

for $k \geq 2$, $A(T_k) = A(Q_4)$.

(This remarkable result is due to Muškardin [22]). Thus, the notion of logical consequence in the k-subject logic (i.e. the logic of T_k) is identical with logical consequence in the Q_4-logic:

for $k \geq 2$, $A \in F$ $C[T_k](A) = C[Q_4](A)$ (15)

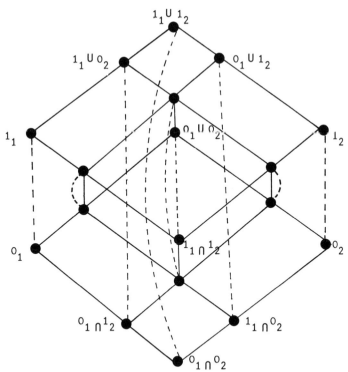

Fig.5. 2-subject truth-values. T_2.

This result can be related to an axiom system of "tautological entailments" due to Anderson and Belnap [2]. They introduced their notion in [1] as the outcome of their search for "first degree entailments" (i.e. entailments between formulas not themselves containing an entailment connective) which do not commit "paradoxes" of relevance. The following axiom schemes (LA) and rules of inference (LR) comprise the **basic system of logical entailment**:

LA0. $0 \vdash A$, $A \vdash 1$
LA1. $A \vdash \neg\neg A$
LA3. $A \ \& \ B \vdash A$
LA4. $A \ \& \ B \vdash B$
LA5. $A \vdash A \ \mathbf{v} \ B$
LA6. $B \vdash A \ \mathbf{v} \ B$
LA7. $A(B \ \mathbf{v} \ C) \vdash AB \ \mathbf{v} \ C$
LR1. from $A \vdash B$ and $B \vdash C$ derive $A \vdash C$
LR2. from $A \vdash B$ derive $\neg B \vdash \neg A$,

LOGIC AND INEXACTNESS

LR3. from $A \vdash B$ and $A \vdash C$ derive $A \vdash BC$,

LR4. from $A \vdash C$ and $B \vdash C$ derive $A \vee B \vdash C$.

Muškardin [23] proved that this system is complete with respect to Q_4 semantics, i.e.

$B \in C[Q_4](A)$ if, and only if, $A \vdash B$.

Thus by (15) the system of tautological entailments axiomatises the notion of logical consequence in the k-subject logic for $k \geq 2$.

* * * * * * * * * *

GENERAL REFERENCES

1. A.R.Anderson & N.D.Belnap Jr., "Tautological Entailments", **Philosophical Studies** 13 (1962) 9-24.
2. A.R.Anderson & N.D.Belnap Jr., **Entailment; the Logic of Relevance and Necessity** Vol.1 (Princeton University Press, 1975).
3. A.Białynicki-Birula & H.Rasiowa, "On the Representation of Quasi-Boolean Algebras", **Bull. Acad. Polon. Science** Cl.III 5 (1957) 259-261.
4. G.Birkhoff, "Lattice Theory", **American Mathematical Society** 1948.
5. U.Blau, **Die Dreiwertige Logik der Sprache** (de Gruyter, 1978).
6. N.Campbell, **Physics - the Elements** (Cambridge, 1920).
7. R.Carnap, **Logical Syntax of Language** (Routledge & Kegan Paul, 1937).
8. R.Cignoli, "Boolean Elements in Łukasiewicz Algebras I", **Proceed. of the Jap. Acad.** 41 (1965) 670-675.
9. J.P.Cleave, "Some Remarks on the Interpretation of Three-valued Logics", **Ratio** XXII(1) (1980) 52-60.
10. J.P.Cleave, "The Notion of Logical Consequence in the Logic of Inexact Predicates", **Zeitschr. für Math. Logik und Grund. der Math.** 20 (1974) 307-324.
11. J.P.Cleave, "Quasi-Boolean Algebras, Empirical Continuity and Three-valued Logic", **Zeitschr. für Math. Logik und Grund. der Math.** 22 (1976) 481-500.
12. J.P.Cleave, "An Account of Entailment Based on Classical Semantics", **Analysis** 34 (1974) 118-122.
13. J.P.Cleave, "The Axiomatisation of Material Necessity", **Notre Dame Journal of Formal Logic** XX (1979) 180-190.

14. G.Grätzer, **Universal Algebra** (Van Nostrand, 1968).
15. P.Hájek, K.Bendová & Z.Renc, "The GUHA Method and the Three-valued Logic", **Kybernetica** 7 (1971) 421-435.
16. J.A.Kalman, "Lattices with Involution", **Trans. Amer. Math. Society** 87 (1958) 458-491.
17. S.C.Kleene, **Introduction to Metamathematics** (North Holland, 1962).
18. K.Kuratowski, **Introduction to Set Theory and Topology** (Pergamon, 1961).
19. Z.Manna & J.McCarthy "Properties of Programs. Partial Function Logic", **Machine Intelligence 5** (eds.) B.Meltzer & D.Mitchie (Edinburgh University Press).
20. G.C.Moisil, "Recherches sur l'Algèbra de la Logique", **Ann. Sci. Univ. Jassy.** 22 (1935) 1-118.
21. A.Monteiro, "Matrices de Morgan Caracteristiques pour le Calcul Propositionnel Classique", **An. Acad. Brasil Ci** 32 (1960) 1-7.
22. V.Muškardin, **Quasi-Boolean Algebras and Semantics for Entailment**, Ph.D.Thesis, University of Bristol, 1978.
23. E.Nagel, **The Structure of Science: Problems in the Logic of Scientific Explanation** (Routledge & Kegan Paul, 1979).
24. H.Rasiowa, **An Algebraic Approach to Non-Clasical Logics** (North Holland, 1974).
25. H.Reichenbach, **Wahrscheinlichkeitslehre** (Leiden, 1935).
26. H.Reichenbach, "Three-valued Logic and the Interpretation of Quantum Mechanics", in **The Logico-Algebraic Approach to Quantum Mechanics** Vol.1 (ed.) C.A.Hooker (D.Reidel, 1975).
27. B.Russell, **Introduction to Mathematical Philosophy** (Allen & Unwin, 1970).
28. K.Schütte, **Beweistheorie** (Springer, 1960).
29. A.Tarski, **Logic, Semantics, Metamathematics** (Oxford, 1956).

* * * * * * * * * *

REFERENCES TO THE WRITINGS OF S.KÖRNER.

K1. "On Entailment", **Proceedings of the Aristotelian Society** XXI (1947) 143-162.
K2. **Conceptual Thinking** (Dover, 1959).
K3. **The Philosophy of Mathematics** (Hutchinson University Library, 1960).

K4. **Experience and Theory** (Routledge & Kegan Paul, 1966).
K5. Private communication to A.Anderson, 29.1.73.
K6. "Material Necessity", **Kant Studien** (1973) 423-430.
K7. **Experience and Conduct** (Cambridge University Press, 1976).
K8. "On Logical Validity and Informal Appropriateness", **Philosophy** 54 (1979) 377-79.
K9. "Science and the Organisation of Belief", in (ed.) D.H.Mellor, **Science, Belief and Behaviour** (Cambridge University Press, 1980).

* * * * * * * * * *

COMMENT ON DR. CLEAVE

Stephan Körner

I am grateful to Dr.Cleave for his systematic formulation of some of my logico-philosophical ideas and for his original contribution, which consists in his showing that my logic of inexactness is related to quasi-Boolean algebra, as classical logic is related to Boolean algebra and intuitionist logic to pseudo-Boolean algebra. That I accept his clarification and improvement of my work is not surprising, since we have for more than twenty years discussed philosophy and logic at regular intervals and since during this period I have never published anything on inexactness or continuity without first seeking his opinion and his general approval.

Our discussion on logico-philosophical questions is still continuing and it is possible that we might not come to a full agreement on the question of alternative logics and their common core. These issues, which are not part of his paper, will be discussed by him in a book on which he is working. They are to some extent discussed by me in my recent book, (**Metaphysics: Its Structure and Function**, Cambridge, 1984).

Although my frequent references to Dr.Cleave's work are sufficient proof of my intellectual indebtedness, I gladly use the present occasion of thanking him explictly for his friendship and help.

* * * * * * * * *

BIBLIOGRAPHY OF

STEPHAN KÖRNER'S WORKS

BOOKS:

Conceptual Thinking (London, Cambridge University Press, 1955; New York, 1959).

Kant (Harmondsworth, Penguin Books, 1979, 10th edition).
    Translations:
        German (Vandenhoeck & Ruprecht, Göttingen, 1967);
        Japanese (1976);
        Spanish (Alianza Editorial, S.A., Madrid, 1977).

The Philosophy of Mathematics (London, Hutchinson University Library, 1960; New York, Dover, 1986).
    Translations:
        Rumanian (Bucuresti, Editura Stiintifica, 1965);
        Spanish (S.A. Gabriel Mancera 65, Mexico, XXI Siglo Editores, 1967);
        German (München, Nymphenburger Verlags-handlung, 1968).

Experience and Theory (London, Routledge & Kegan Paul, 1966; 2nd ed., 1969; New York, The Humanities Press, 1966).
    Translations:
        Rumanian (Bucuresti, Editura Stiintifica, 1969);
        Czech (Praha, Nakladatelstvi Svoboda, 1970);
        German (Frankfurt am Main, Suhrkamp Verlag, 1970).

Fundamental Questions in Philosophy (London, Penguin Press, 1969; 4th ed., London, Harvester, 1980).
    Translations:
        German (München, Paul List Verlag KG, 1970);
        Spanish (Barcelona, Editorial Ariel, 1976).
Earlier published under the title:
What is Philosophy? - One Philosopher's Answer.

Categorial Frameworks (Oxford, Blackwell, 1970);
    Translation:
        Italian (Petrinelsi, Milano, 1983).

Experience and Conduct (London, Cambridge University Press, 1976;

paperback 1980).

Metaphysics: Its Structure and Function (London, Cambridge University Press, 1984; paperback 1986).

EDITOR OF:

Observation and Interpretation in the Philosophy of Physics, edited in collaboration with M.H.L.Pryce (paperback, London Dover, 1957);

Practical Reason, (Oxford, Blackwell, and Yale University Press, 1974);

Explanation, (Oxford, Blackwell, and Yale University Press, 1975).

Philosophy of Logic, (Oxford, Blackwell, and California University Press, 1976).

Selected Papers of Abraham Robinson, edited with H.J.Keisler, W.A.J.Luxemburg and A.D.Young (Yale University Press, 1979). Introduction to Vol.II, pp.xli-xlv.
F.Brentano, Raum, Zeit und Kontinuum, edited with R.M.Chisholm (Hamburg, Felix Meiner, 1976).

PAPERS:

1946-47   "On Entailment", **Proceedings of the Aristotelian Society.**

1948   "The Nature of Some Metaphysical Propositions", **Mind;**
"Are All Philosophical Questions Questions of Language?" **Proceedings of the Aristotelian Society,** Supplementary Volume.

1950   "Entailment and the Meaning of Words", **Analysis.**

1951   "Ostensive Predicates", **Mind.**
"On Some Moral and Other Concepts", **Philosophy and Phenomenological Research.**

1951-52   "On Theoretical and Practical Appropriateness", **Proceedings of the Aristotelian Society** 52, 119-34.

1953   "On Laws of Nature", **Mind** 62, 216-29.
"The Nature of Philosophical Analysis", **Proceedings of the IX International Congress of Philosophy** 5, 121-26.
"The Notion of Infinity", **Proceedings of the Aristotelian Society,** Suplementary Volume, 53-68.

1954   "Individuals and Properties", **Mind,** 379-83.

1955   "Truth as a Predicate", **Analysis** 15, 106-10.

1957   "Types of Philosophical Thinking", **British Philosophy in Mid-**

Century, (ed.) C.A.Mace (London, Allen & Unwin, 1957) Ch.4.
"Reference, Vagueness and Necessity", **Philosophical Review** 66, 363-77.
"Philosophical Arguments in Physics", in **Observation and Interpretation**, (ed.) S.Körner in collaboration with M.H.L.Pryce, **Colston Papers** Vol.9 (London, Butterworth) 97-102.
"Some Remarks on Philosophical Analysis", **Journal of Philosophy** 54, 758-66.

1959 "Determinables and the Notion of Resemblance", **Proceedings of the Aristotelian Society**, Supplementary Volume 53, 125-41.
"Über reine und angewandte Mathematik", **Ratio**, 19-35.
"The Relation Between Exact and Inexact Concepts", **Proceedings of the XIII International Congress of Philosophy** (Venice, 1958).
"Watkins on Metaphysics", **Mind** 68, 548-49.
"The Nature of Pure and Applied Mathematics", (English version) **Ratio**.

1960 "Philosophical Arguments in Physics", in **The Structure of Scientific Thought**, (ed.) E.Madden (London, Routledge & Kegan Paul) 106-11.
"Philosophical Method and C.D.Broad's Philosophy", in **The Philosophy of C.D.Broad**, (ed.) P.Schilpp, Library of Living Philosophers (New York, 1959) 95-114.
"Kant", in **Western Philosophy and Philosophers**, (ed.) J.Urmson (London, Hutchinson).

1961 "On the Relevance of Philosophy to Mathematics and the Sciences", **Yale Scientific Magazine** 35, 12-17.
"Neo-Kantianism", **Encyclopaedia Britannica**.

1962 "Philosophy, Science and Commonsense", **Annals of the Japan Association for Philosophy of Science** 2, 114-19; also in Japanese, as part of book.
"On Empirical Continuity", **Monist** 47, 1-19.

1963 "Anderson's Philosophy of Experience", **Quadrant** 7.

1964 "On Making Room for Faith: A Humanist's View", in **Theology and the University** (London, Longman & Todd) Ch.6.
"The Concept of Coherence", in **Philosophical Interrogations**, (ed.) Sydney & Beatrice Rome (New York, Holt, Rinehart & Winston) Ch.4.
"Deductive Unification and Idealization", **British Journal for the Philosophy of Science**, 274-84.
"Substance", **Proceedings of the Aristotelian Society**, Supplementary Volume, 79-90.
"Science and Moral Responsibility", **Mind**, 161-72.
"An Empiricist Justification of Mathematics", **Proceedings of the International Congress for Logic, Methodology and Philosophy of Science**, Jerusalem.

1966 Introduction to G.C.Field's **Moral Theory** (Methuen, University Paperbacks) ix-xxiii.
"On Deductivism as a Philosophy of Science", in **Metaphysics and**

Explanation, (ed.) W.H.Capitan & D.D.Merrill (Pittsburgh).
"Zur Kantischen Begründung der Mathematik und der Naturwissenschaften", **Kant-Studien** 56, 463-73.
"On the Concept of the Practicable", Presidential Address, **Proceedings of the Aristotelian Society**, 1-16.
"Transcendental Tendencies in Recent Philosophy", **Journal of Philosophy** 63, 551-61.
"On the Logical Function and Structure of Scientific Theories", **Science Progress** 54, 1-12.
"Some Relations Between Philosophical and Scientific Theories", Presidential Address, **British Journal for the Philosophy of Science** 17.
"On the Concept of Truth in Western Philosophy", Public Lecture, **Visva-Bharati Journal of Philosophy** 3.
Review articles:
"The Philosophy of Nelson", **Journal of Philosophy** 63, 782-94;
"The Philosophy of Carnap", **Mind**, 285-93;
"Science and Moral Responsibility", **Mind**.

1967
"The Implications of Post-Gödelian Mathematics", in **Problems in the Philosophy of Mathematics**, (ed.) I.Lakatos (Amsterdam) 118-38.
"A New Neo-Kantianism", **Times Literary Supplement**, 1 June.
"Continuum", "Laws of Thought", "Cassirer", **Encyclopaedia of Philosophy** (ed.) P.Edwards (New York, Collier-Macmillan).
"The Impossibility of Transcendental Deductions", **Monist** 51, 317-31.
"Kant's Conception of Freedom", Dawes Hicks Lecture on Philosophy, **Proceedings of the British Academy** 53.

1968
Foreword to **Contemporary Science and Realism** by R.Blanche (Edinburgh, Oliver & Boyd).
"Mathematical Frameworks in Scientific Thinking", **Advancement of Science**, March, 306-10.
"On Bergman's Ontology", **Philosophy of Science** 35, 64-71.
"Reply to Mr. Kumar", **British Journal for the Philosophy of Science** 18, 323-24.

1969
"On Fitting Theories to the World: A Comment", **Philosophy Forum** 7, 78-85.
"Categorial Change and Philosophical Argument", **Proceedings of the Israel Academy of Science and Humanities** 3.
"Existence Assumptions in Practical Thinking", in **Fact and Existence**, (ed.) J.Margolis (Oxford, Blackwell, 1969).
"Reply to Mr. Harre´", **Mind** 78, new series.
"Extra-mural Experience", **Adult Ed.** 41.

1970
"Description, Analysis and Metaphysics", in **The Nature of Philosophical Inquiry**, (ed.) Joseph Bobik (South Bend, Indiana, Notre Dame University Press) Ch.1.
"On the Kantian Foundations of Science and Mathematics", in **The First Critique - Reflections on Kant's Critique of Pure Reason**, (ed.) T.Penelhum & J.J.MacIntosh (California, Wadsworth).
"On the Foundations of Mathematics in Experience", **L'Age de la Science** 3, 187-200.

BIBLIOGRAPHY

Various items in **Historisches Wörterbuch der Philosophie**, (ed.) Joachim Ritter (Stuttgart, Basel).
"Prolegomena to the 4th International Congress of Logic, Methodology and Philosophy of Science, Bucharest", **Forum** 13, 6-12.
Editor with Paul Bernays and others of Volumes, 1,5,6 and 7 of Leonard Nelson's **Gesammelte Werke** (Hamburg, Meiner).
"Abstraction in Science and Morals", 24th Eddington Memorial Lecture (Cambridge University Press).

1972 "On a Difference Between the Natural Sciences and History", in **Biology, History and Natural Philosophy**, (ed.) A.D.Breck & W.Yourgrau (New York, Plenum) 243-63.
"Mathematik als Wissenschaft formaler Systeme", in **Grundlagen der modernen Mathematik**, (ed.) H.Meschkowski (Darmstadt, Wissch Buchgesellschaft) 124-57.
"Logic and Conceptual Change", in **Conceptual Change**, (ed.) G.Pearce & P.Maynard (Dordrecht, Reidel).
"Individuals in Possible Worlds", in **Logic and Ontology**, (ed.) M.Munitz (New York, New York University Press).
"On the Coherence of Factual Beliefs and Practical Attitudes", **American Philosophical Quarterly** 9, 1-18.
Editor with Paul Bernays and others of Volumes 4 & 9 of Leonard Nelson's **Gesammelte Werke** (Hamburg, Meiner).

1973 "Rational Choice", Presidential Address, **Proceedings of the Aristotelian Society**, Supplementary Volume, 1-18.
"Necessity in History and Philosophy", **Dictionary of the History of Ideas**, Vol.3 (New York) 351-62.
"Classification Theory", Encyclopaedia Brittanica, Vol.1, 691-94.
Editor with Paul Bernays and others of Volume 2 of Leonard Nelson's **Gesammelte Werke** (Hamburg, Meiner).
"Material Necessity", **Kant-Studien** 64, 423-30.
"The Claims of Metaphysics", review of A.J.Ayer's **The Central Questions of Philosophy, New Statesman**, December.

1974 "On the Structure of Codes of Conduct", **Mind**, January, 61-75.
Editor with Paul Bernays and others of Volume 9 of Leonard Nelson's **Gesammelte Werke** (Hamburg, Meiner).
"Theoretische und praktische Vernunft", **Akten des 4 Internationalen Kant Kongresses**, Mainz, 6-10 April.
"Empiricism in Ethics", in **Impressions of Empiricism**, Royal Institute of Philosophy Lectures Vol.9, 174-92.

1975 "Classical Logic and Inexact Predicates: A Reply", **Mind** 84.
"On Some Relations Between Logic and Metaphysics", in **The Logical Enterprise**, (ed.) Alan R.Anderson et.al. (New Haven, Yale University Press).
"On the Identification of Agents", **Philosophia: Philosophy Quarterly of Israel** 5 (January-April).
"Philosophieren in der Gegenwart", **Evangelische Kommentare**, (August).
"On the Relevance of Philosophy", The First Norman Liddiard Memorial Lecture delivered in Swindon in 1976 (Swindon, Waterleaf Press).
"Vagueness in the Language of Science", **Dialogue** 14, 306-30.

BIBLIOGRAPHY

1976   "Classification Theory", **International Classification** 3.
"A Humanist's Reflections on Morality, Religion and the Churches", **Epworth Review** 3, 71-77.
"On Abraham Robinson's Philosophy of Mathematics", **Bulletin of London Mathematical Society** 8, 316-19.
Foreword to **Philosophische Untersuchungen zu Raum, Zeit und Kontinuum** by Franz Brentano, (ed.) S.Körner & R.M.Chisholm.
"On the Logic of Relations", **Proceedings of the Aristotelian Society**, n.s. 77.
"On the Subject Matter of Philosophy", in **Contemporary British Philosophy** (ed.) H.D.Lewis (London, Allen & Unwin, 1976).
"Über die Voraussetzungen der Transcendental philosophie", **Grazer Philosophische Studien** 2, 214-20.
"Kant's Conception of Freedom", Dawes Hicks Lecture (The British Academy).

1977   "Logik und Begriffswandel", Sonderdruck aus **Metaphysik**, 404-22.
"Über ontologische Notwendigkeit und die Begründung ontologischer Prinzipien", **Neue Hefte für Philosophie**, Heft 14 (Zur Zunkunft der Transzendental-philosophie).
"Zur Immanenten und Transzendenten Metaphysik", **Perspektiven der Philosophie**, Band 4.

1978   "Über Brentanos Reismus und die Extensionale Logik", **Grazer Philosophische Studien** 5.

1979   "Ayer on Metaphysics", in **Perception and Identity**, (ed.) G.K.Macdonald (London, Macmillan) Ch.12, 262-76.
"Leonard Nelson und der philosophische Kritizismus", **Vernunft, Erkenntnis und Sittlichkeit**, (ed.) P.Schröder (Hamburg, Meiner) Ch.1, 1-17.
"On Russell's Critique of Leibniz's Philosophy", **Bertrand Russell Memorial Volume**, (ed.) George W.Roberts (Allen & Unwin).
"On Bennett's so-called Analytic Transcendental Arguments", in **Transcendental Arguments and Science**, (ed.) P.Bieri et.al. (Dordrecht, Reidel), 65-71.
"On Logical Validity and Informal Appropriateness", **Philosophy** 54, 377-79.

1980   "Science and the Organization of Belief", in **Science, Belief and Behaviour**, (ed.) D.H.Mellor (London, Cambridge University Press), 43-59.
Review of Gordon G.Brittan Jr., **Kant's Theory of Science**, in **Synthese** 45, 311-15.
"Wissenschaft", in **Handbuch wissenschaftstheoretischer Begriffe** (Göttingen, Vandenhoeck & Ruprecht) 726-37.

1981   "On the Nature of Limits of Cognitive and Evaluative Relativity" in **200 Jahre Kritik der reinen Vernunft**, (ed.) J.Kopper & W.Marx (Gerstenberg Verlag, Hildesheim) 113-128.
"Wittgenstein und die philosophische Tradition", Inaugural Lecture in **Ethik-Grundlagen, Probleme und Anwendungen, Akten des 5. Internationalen Wittgenstein Symposiums 25. bis 31 August 1980, Kirchberg am Wechsel (Österreich)** (Vienna, Morscher &

Stranzinger) 431-33.
"Über Sprachspiele und rechtliche Institutionen" (ibidem) 480-491.

1982 Reply to Bertil Rolf's "Körner on Vagueness and Applied Mathematics" in **Grazer Philosophische Studien** 15, 109-119.
"On the Empirical Application of Mathematics and Some of its Philosophical Aspects" in **Proceedings of the Israel Colloquium for History, Philosophy and Sociology of Science**, 1-14.

1983 "Thinking, Thought and Categories" in **Monist** (July).
"On Kant's and Some Post-Kantian Conceptions of A Priori Knowledge" in **New Trends in Philosophy** (Tel Aviv) 185-204.
Also to be published by Humanities Press.

1984 "Über philosophische Methoden und Argumente", **Grazer Philosophische Studien** 22, 27-39.

1986 "Scientific Information, Explanation and Progress" (Inaugural Address) in **Proceedings of the 7th International Congress of Logic, Methodology and Philosophy of Science** (ed.) Ruth B. Marcus et.al. (North-Holland) 1-17.

TO APPEAR:

"On Brentano's Objections to Kant's Philosophy" in **Topoi**.
"Some Clarifications and Replies" in **Grazer Philosophische Studien**.
"On Kant's Theory of Concepts" in **Acts of the 6th International Kant Congress in the State University of Pennsylvania**.
"On Some Methods and Results of Philosophical Analysis" in **Philosophy in the U.K. Today**, (ed.) G.S. Shanker (London, Croom Helm) 154-170.

* * * * * * * * * *

INDEX OF NAMES

ALLISON, Henry E., 51
ANDERSON, A.R., 151,156-7,159
ANSCOMBE, G.E.M., 106
ARISTOTLE, 5,12,30,35,41,47,70,73,
  76-7,80,85-6,89,95,96,100-1,
  104-5,111,116
AUSTIN, J.L., 93,105
AVICENNA, 115

BECK, L.W., ix, 52-7,132,134
BELNAP, N.D.Jr., 151,156-7
BENDOVA, K., 158
BERKELEY, George, 66
BIAŁYNICKI-BIRULA, A., 146,157
BIRKHOFF, G., 149,157
BLANSHARD, Brand, 85-6
BLAU, U., 157
BOLZANO, B., 80,141-2
BRADLEY, F.H., 80,96
BRENTANO, Franz, 1-2,5-15,30,64,71
BUTLER, J., 89,92,109-11

CAMPBELL, N., 138,157
CARNAP, R., 28,139,157
CHARLES, David, 106
CHISHOLM, R.M., 9-11,13,15,87
CIGNOLI, R., 157
CLEAVE, J.P., 30,33,157,160

COCKBURN, David, 106
COLLINGWOOD, R.G., 70
DAVIDSON, Donald, 78,81-3,93,95-7,102-3,
  105
DESCARTES, Rene, 1,9-10,71,82,85-7

EDDINGTON, Arthur Stanley, 131
EDWARDS, Paul, 131
EINSTEIN, Albert, 10
EUCLID, 31,152

FODOR, J.A., 21-3,25-6
FREEMAN, Frances, ix
FREEMAN, Samantha, ix
FREGE, Gottlob, 61,64,116-17,132
FREUD, Sigmund, 94

GEACH, P.T., 154
GENTZEN, Gerhardt, 143-4
GOETHE, 55
GRÄTZER, G., 158

HAJEK, P., 140,158
HARE, R., 105
HEGEL, G.W.F., 78-80,85-7
HILBERT, D., 143-4
HOCHBERG, H., 81-2
HOLBACH, Baron, 128,131

INDEX

HOLOWKA, J., 132-4
HOOK, Sidney, 131
HUME, David, 90
HUSSERL, Edmund, 64,71
HUXLEY, A., 81

IMBERGER, H., ix
INGARDEN, Roman, 63-4,67,73
ISRAELI, Isaac, 115

KALMAN, J.A., 146,158
KANT, Immanuel, 10,12,14,28,35-53, 55-7,70-71,73-4,86-7,118
KASTIL, Alfred, 8,15
KENNY, Anthony, 99,105
KLEENE, S.C., 30,33,140,146,158
KÖRNER, S., 1,3-4,7,17-20,22,24-6,35, 49-51,59-61,64-8,75,77-82,89-92,94, 98-101,104,113-5,131,137-40,147-53
KOTARBINSKI, Tadeusz, 61
KUNG, Guido, 64-5
KURATOWSKI, K., 158

LANDE, Alfred, 119,131
LEHRER, Keith, 26-29,32
LEIBNIZ, Gottfried Wilhelm, 9,53, 63,71,73,77,113,132-4
LESNIEWSKI, S., 66-7
LEWIS, C.I., 151
LEWY, C., 154
LIDDELL, H.G., 93
LOCKE, John, 53-4

McCARTHY, J., 158
MANNA, Z., 158

MARCISZEWSKI, W., 69-70, 72-4, 111
MEINONG, 1-2,5-7,10,76-7,82
MOISIL, G.C., 146,149,158
MONTEIRO, A., 146,158
MORTON, Adam, ix
MUŠKARDIN, V., 141,154-5,157-8

NAGEL, E., 137,158
NEWTON, Sir Isaac, 10,54,56,58,134

PAP, Arthur, 131
PAUL (Saint), 94
PEARS, David, 93,105-6
ANDERSON, Stig A., 67-8
PIERCE, Charles S., 54
PIGDEN, Charles, 80
PLATO, 11,31,64,105
PLUMWOOD, V., 82
PYTHAGORUS, 152

QUINE, W.V.O., 63,66

RASIOWA, H., 146,157-8
REID, Thomas, 17,21-6,32
REICHENBACH, H., 138,158
RENC, Z., 158
ROUTLEY (see SYLVAN)
RUSSELL, Bertrand, 64-5,81,158
RYLE, Gilbert, 65

SANTAYANA, G., 104
SARTRE, Jean-Paul, 39,57
SCHLICK, Moritz, 2,4-5,7
SCHULZ, J., 39
SCHUTTE, K., 152,158

INDEX

SCOTT, D., 93
SELLARS, W., 19-21,25-6
SHARPE, Robert, 107,109-11
SMART, J.J.C., 82-3
SMILEY, T., 154
SMITH, J.C., 26
SOCRATES, 97
SYLVAN, Richard, 82,84-6

TARSKI, Alfred, 75-6,80,82,84-5, 141-2,158
TOMIN, Julius, 106

TWARDOWSKI, K., 64

von NEUMANN, J., 71

WALSH, Dorothy, 100
WEBSTER, 64
WEINGARTNER, Paul, 62,67
WESTFALL, Richard S., 58
WHITEHEAD, Alfred North, 77
WINTHER, Josie, ix
WITTGENSTEIN, Ludwig, 31,81,83

NIJHOFF INTERNATIONAL PHILOSOPHY SERIES

Applying Philosophy

1. Rotenstreich N: Philosophy, History and Politics – Studies in Contemporary English Philosophy of History. 1976. ISBN 90-247-1743-4.
2. Srzednicki JTJ: Elements of Social and Political Philosophy. 1976. ISBN 90-247-1744-2.
3. Tatarkiewicz W: Analysis of Happiness. 1976. ISBN 90-247-1807-4.
4. Twardowski K: On the Content and Object of Presentations – A Psychological Investigation. Translated and with an Introduction by R Grossman. 1977. ISBN 90-247-1726-7.
5. Tatarkiewicz W: A History of Six Ideas – An Essay in Aesthetics. 1980. ISBN 90-247-2233-0.
6. Noonan HW: Objects and Identity – An Examination of the Relative Identity Thesis and Its Consequences. 1980. ISBN 90-247-2292-6.
7. Crocker L: Positive Liberty – An Essay in Normative Political Philosophy. 1980. ISBN 90-247-2291-8.
8. Brentano F: The Theory of Categories. Translated by RM Chisholm and N Guterman. 1981. ISBN 90-247-2302-7.
11. Hoffman P: The Anatomy of Idealism – Passivity and Activity in Kant, Hegel and Marx. 1982. ISBN 90-247-2708-1.
12. Gram MS: Direct Realism – A Study of Perception. 1983. ISBN 90-247-2870-3.
14. Smith JW: Reductionism and Cultural Being – A Philosophical Critique of Sociobiological Reductionism and Physicalist Scientific Unificationism. 1984. ISBN 90-247-2884-3.
15. Zumbach C: The Transcendent Science – Kant's Conception of Biological Methodology. 1984. ISBN 90-247-2904-1.
16. Notturno MA: Objectivity, Rationality and the Third Realm: Justification and the Grounds of Psychologism – A Study of Frege and Popper. 1984. ISBN 90-247-2956-4.
18. Russell JJ: Analysis and Dialectic – Studies in the Logic of Foundation Problems. 1984. ISBN 90-247-2990-4.
20. Broad CD: Ethics. Edited by C Lewy. 1985. ISBN 90-247-3088-0.
21. Seargent DAJ: Plurality and Continuity – An Essay in GF Stout's Theory of Universals. 1985. ISBN 90-247-3185-2.
22. Atwell JE: Ends and Principles in Kant's Moral Thought. 1986. ISBN 90-247-3167-4.
29. Brentano F: On the Existence of God – Lectures given at the Universities of Würzburg and Vienna (1868–1891). 1987. ISBN 90-247-3538-6.
33. Young J: Willing and Unwilling: A Study in the Philosophy of Arthur Schopenhauer. 1987. ISBN 90-247-3556-4.

Logic and Applying Logic

9. Marciszewski W (ed): Dictionary of Logic as Applied in the Study of Language – Concepts / Methods / Theories. 1981. ISBN 90-247-2123-7.
10. Ruzsa I: Modal Logic with Descriptions. 1981. ISBN 90-247-2473-2.
13. Srzednicki JTJ and Rickey VF (eds): Leśniewski's Systems – Ontology and Mereology. ISBN 90-247-2879-7.
24. Srzednicki JTJ and Stachniak Z (eds): S. Lesniewski's Lecture Notes in Logic. 1987. ISBN 90-247-3416-9.

Contributions to Philosophy

17. Dilman I (ed): Philosophy and Life – Essays on John Wisdom. 1984. ISBN 90-247-2996-3.
19. Currie G and Musgrave A (eds): Popper and the Human Sciences. 1985. ISBN 90-247-2998-X.
25. Taylor BM (ed): Michael Dummett – Contributions to Philosophy. 1987. ISBN 90-247-3463-0.
28. Srzednicki JTJ (ed): Stephan Körner – Philosophical Analysis and Reconstruction. 1987. ISBN 90-247-3543-2.

Main Stream

23. Agassi J and Jarvie IC (eds): Rationality: The Critical View. 1987. ISBN 90-247-3275-1.
26. Bar-On AZ: The Categories and Principle of Coherence – Whitehead's Theory of Categories in Historical Perspective. 1987. ISBN 90-247-3478-9.

ABA-4011

WITHDRAWN FROM
THE ELLEN CLARKE BERTRAND LIBRARY